The Patrick Moore Practical Ast

CW01645328

For further volumes:
http://www.springer.com/series/3192

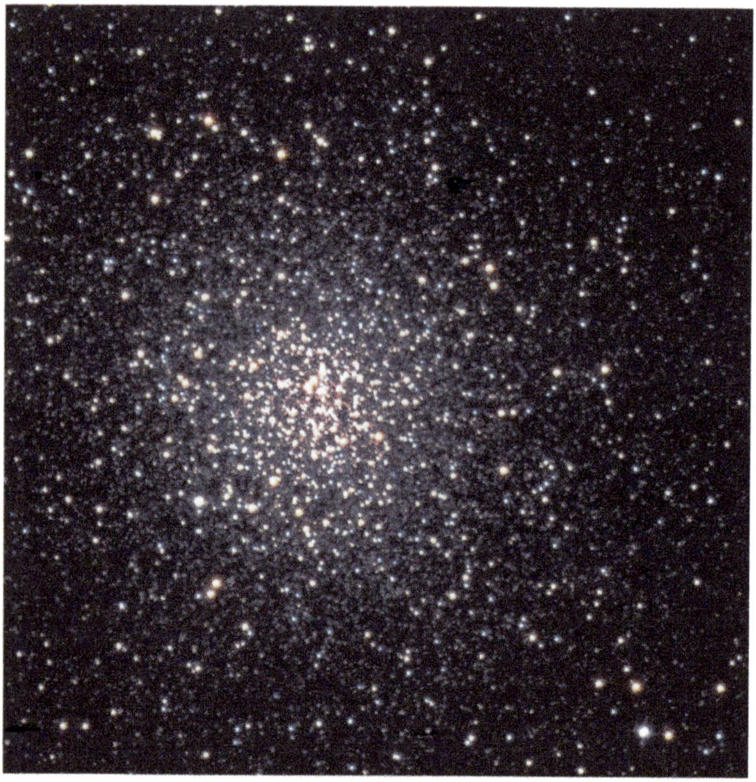

Globular cluster Messier 5 in Serpens. Image by Martin Griffiths courtesy of LCOGT

Choosing and Using Astronomical Filters

Martin Griffiths

Springer

Martin Griffiths
University of South Wales
Glyntaf, UK

ISSN 1431-9756 ISSN 2197-6562 (electronic)
ISBN 978-1-4939-1043-4 ISBN 978-1-4939-1044-1 (eBook)
DOI 10.1007/978-1-4939-1044-1
Springer New York Heidelberg Dordrecht London

Library of Congress Control Number: 2014940529

© Springer Science+Business Media New York 2015
This work is subject to copyright. All rights are reserved by the Publisher, whether the whole or part of the material is concerned, specifically the rights of translation, reprinting, reuse of illustrations, recitation, broadcasting, reproduction on microfilms or in any other physical way, and transmission or information storage and retrieval, electronic adaptation, computer software, or by similar or dissimilar methodology now known or hereafter developed. Exempted from this legal reservation are brief excerpts in connection with reviews or scholarly analysis or material supplied specifically for the purpose of being entered and executed on a computer system, for exclusive use by the purchaser of the work. Duplication of this publication or parts thereof is permitted only under the provisions of the Copyright Law of the Publisher's location, in its current version, and permission for use must always be obtained from Springer. Permissions for use may be obtained through RightsLink at the Copyright Clearance Center. Violations are liable to prosecution under the respective Copyright Law.
The use of general descriptive names, registered names, trademarks, service marks, etc. in this publication does not imply, even in the absence of a specific statement, that such names are exempt from the relevant protective laws and regulations and therefore free for general use.
While the advice and information in this book are believed to be true and accurate at the date of publication, neither the authors nor the editors nor the publisher can accept any legal responsibility for any errors or omissions that may be made. The publisher makes no warranty, express or implied, with respect to the material contained herein.

Cover illustration: The Carina Nebula photograph, credit: NASA, ESA, and M. Livio and the Hubble 20th Anniversary Team (STScI).

Printed on acid-free paper

Springer is part of Springer Science+Business Media (www.springer.com)

This book is dedicated to:
Dena, Harry, Yoda and Gloria
All of them are stars in their own special way…

About the Author

Martin Griffiths is an enthusiastic science communicator, writer and professional astronomer. Over his career he has utilized history, astronomy and science fiction as tools to encourage greater public understanding of science. He is a recipient of the Astrobiology Society of Britain's Public Outreach Award (2008) and the Astronomical League's Outreach Master Award (2010). He also holds the League's Master Observer certificate and has written or contributed to over 100 published science articles for many journals. He is currently an astronomer and senior lecturer at the University of South Wales in the UK.

He was one of the founding members of NASA's Astrobiology Institute Science Communication Group, which was active in 2003–2006. He also managed a large ESF program in Astrobiology for adult learners across Wales in 2003–2008. Since then he has been involved in promoting adult education, assisting in the development of a new observational astronomy award at the University of South Wales, and various other projects, including initial consultation on the setup of an educational observatory in Andalucia, Spain, now run by Andy Burns and Kath Griffiths.

He is a consultant to the Welsh Government through his involvement with the Dark Sky Discovery initiative, enabling public access to dark sky sites in association with Dark Sky Wales, Dark Sky Scotland and Natural England. He was also responsible for surveying the sky quality of the Brecon Beacons National Park in their successful bid to gain International Dark Sky Association Dark Sky Reserve status in 2013 and is a consultant to the Hay Tourism Board for their annual dark sky festivals.

Martin is a Fellow of the Royal Astronomical Society; a Fellow of the Higher Education Academy and a member of the Astrobiology Society of Britain; the European Society for the History of Science; the British Astronomical Association; the British Science Association; the Webb Deep-Sky Society; the Society for Popular Astronomy and the Astronomical League. Martin is also a local representative for the BAA Campaign for Dark Skies and lectures in astronomy to anyone who will listen.

Preface

Taking a quality image of the night sky used to be a daunting task for any photographer and was a valid achievement in its own right. Capturing the beauty of the astronomical objects involved real skills.

In the past, achieving this was always something that engendered pride in the photographer and the admiration of peers. But with the advent of inexpensive digital SLR and CCD cameras, achieving good quality images is now within reach for all astronomers whatever their experience is and is relatively simple to accomplish. And, there are filters that can help them get even better pictures.

Alternatively, there are many amateur astronomers out there who still prefer to observe astronomical objects directly but use filters to enhance their visual acuity and render the objects visible via the medium of sketching. Coloured filters that enhance the visibility of features in planetary atmospheres and surfaces are still commercially available and are regularly offered with some telescopes. "Moon" filters and "solar" filters are still making an impact on new generations of observers who wish to know the sky in as much visual detail as possible.

The manufacturing and supply of filters for astrophotography and visual observing is now a large industry. The purpose of this book is to introduce these filters and give some basic advice on their use and application. This book will serve as a reference point for the observer who wishes to gain experience in CCD or DSLR imaging or in simply looking for faint detail in heavenly objects.

A wealth of material exists on the Internet, from the purchasing of filters through to information on the properties of various astronomical filters. By making these disparate items of information easily accessible in one handy book, the author hopes that the matter conveyed here will be a ready guide to the application and use of filters in visual astronomy and astrophotography.

This work cannot by its very nature as a book keep up with everything available in the market, as new technology or enhancement of current materials goes on continuously. However, it will be a useful source for how to use and properly apply filters to the task at hand. In addition, we have supplied a tempting list of objects to observe or image with such filters. This selection of Astronomical Objects takes into account both the size of the observer's telescope and his or her geographical location, as many Messier objects are included and bright NGC objects across both the Northern and Southern Hemisphere.

The photographs, unless otherwise indicated, were taken by myself.

I hope that you find this book to be a useful addition to your astronomical library.

Glyntaf, United Kingdom Martin Griffiths
2014

Acknowledgements

When I started observing and photographing the night sky, I never thought that many of the things I had seen and imaged could become the focus of a book. As a professional astronomer who started as an amateur observer, I am aware of the importance of filters in astrophotography and visual work. It was a chance conversation with Dr Mike Inglis at Springer that became the foundation of this book and I thank him for his trust and support.

I would like to thank the Las Cumbres Global Observatory Global Telescope (LCOGT) and the Faulkes telescope team for their assistance and access to the Faulkes telescopes and archive to provide the images, especially Dr Ed Gomez and Professor Paul Roche for their help and credit for the images. I would also like to thank Nick Howes and Andy Burns of the Wiltshire Astronomical Society for their contributions in supplying advice and some of the images. I would also like to thank the editorial staff at Springer for their helpful comments and suggestions.

Unless otherwise acknowledged in the text, all photographs have been taken by the author and are copyright as are the images of the LCOGT and Faulkes telescope archive and those of Nick Howes and Andy Burns.

Contents

Chapter 1

A Brief History of Astronomical Imaging

The activity of photographing and imaging the objects in the sky has an illustrious but very short history. For centuries after the invention of the telescope astronomers relied upon their eyes to record astronomical details, with wildly varying results.

The problem with astronomical observing at the time of the early nineteenth century was simply this: large telescopes were readily available, but recording any image required an astronomer at the eyepiece who would then sketch the image visible through the telescope.

This technique inevitably led to differences in appearance of the same objects drawn by various observers, since the size of the telescopes may have differed, as well as seeing conditions and elevation of the object. Above all, the dependence on the artistic ability of the astronomer rendered objects liable to disparity in appearance due to the diverse ability of the observer's hand–eye–brain coordination and a tendency to over-exaggerate the image with details that were barely visible, yet rendered out of proportion to their actual values. It was essential that a standard tool be found to record astronomical images as faithfully as possible.

From the first tentative steps into photography made by pioneers such as Louis Daguerre, William Fox Talbot and John Herschel it was realized that is this process could be improved and perfected, that it would make a very useful technological tool for astronomers looking for an absolute standard in observatory work. However, most photographic processes during the early nineteenth century were too slow for astronomical purposes. Daguerre's original process was painstakingly slow, with a polished silver layer on a copper base being exposed to iodine vapor and thus creating an image of silver iodide that then had to undergo further processing. William Fox Talbot made a valuable contribution by substituting the copper base for paper. It was John Herschel's seminal paper, however, "Note on the art of Photography, or The Application of the Chemical Rays of Light to the Purpose of

M. Griffiths, *Choosing and Using Astronomical Filters*, The Patrick Moore Practical Astronomy Series, DOI 10.1007/978-1-4939-1044-1_1,

© Springer Science+Business Media New York 2015

Pictorial Representation," to the Royal Society on March 14, 1839, that made everyone sit up and explore the possibilities of this new technology.

Astronomical application followed shortly thereafter. The first daguerreotype of the Moon was obtained in 1840 by the American J. W. Draper in a 20-min exposure (the plate was later destroyed in a fire), and the Sun was photographed by Leon Foucault in 1845. The first star to be recorded photographically was captured in 1850 by the astronomer William Cranch Bond at Harvard, with the star Vega as his subject, and in 1851 the Sun's corona was photographed for the first time by August Busch at Konigsberg observatory. Nevertheless, it was obvious that the photographic process needed a boost to obtain faster speeds.

A year after Bond took his starry photograph; the French inventor Gustav le Grey invented the wet collodion process that remarkably increased the speed and efficiency of photographic emulsions. The problem was that the film had to be prepared very carefully and used immediately, as the materials could not be stored. This would have led to observatories becoming large photographic shops in their own right by using chemicals such as sulfuric acid and potassium nitrate, which were then reacted on a small quantity of cotton to create nitrocellulose. The whole was then mixed with alcohol, bromide, cadmium, ammonia and iodine and then spread onto glass plates for immediate use. The plate was then developed in silver nitrates and silver iodides, making the whole process a nightmare of preparation requiring the use of noxious chemicals by a large staff specially trained in this type of work.

Unsurprisingly, the wet collodion process did not catch on in astronomy despite being suggested for this use by the British photographer Frederick Scott-Archer. The system was improved considerably by the Welsh landowner John Dillwyn Llewellyn in 1856, who invented the Oxymel process that enabled the plates to be stored for longer, and it was with this improved system that he and his daughter Thereza Took the first photograph of the Moon from Wales. Cementing the relationship with astronomy, Thereza later married the grandson of Neville Maskelyne, the fifth Astronomer Royal, although her astronomical pursuits did not continue after her marriage.

The collodion process did have some success under the application of Warren de la Rue, who took several exquisitely detailed pictures of the Moon during the years 1851–1856. de la Rue's great contribution to celestial photography came with his invention of the photoheliograph, which he presented to the Royal Society. This device enabled him to obtain photographs of the Sun, and in 1858 regular work photographing the Sun was inaugurated at Kew before the process was moved to the Royal Observatory at Greenwich in 1873. In 1860 de la Rue took the photoheliograph to Spain to view the solar eclipse of that year, and the resultant photographs proved beyond doubt that the prominences seen during former eclipses were definitely associated with activity on the solar surface.

Warren de la Rue's contribution to celestial photography in the nineteenth century put imaging at the forefront of astronomical techniques, though by 1873 he had given up photography due to ill health. He donated his telescopes and instruments to the Oxford college observatory and ensured the success of astronomical photography by donating a 13-in. refractor to the college to enable it to take part in the international photographic survey of the heavens that provided data for the French astrographic catalog detailed below. Thanks to his wonderful contributions to this emerging craft, a crater on the Moon is named in his honor.

Advancing Technology

A better dry process was invented that enabled photographers and astronomers to store the plates until ready for use. In 1871 the British inventor Richard L. Maddox produced a dry process using silver salts, and 4 years later J. Johnson and W. Bolton made the first true negative emulsion for photographic use. This system, using the first good photographic quality dry plates, was used for astronomical purposes by the husband and wife team Sir William and Lady Margaret Huggins in 1876, though Huggins and his assistant, the chemist William Miller, had used the wet collodion process to obtain spectrograms of the stars Sirius and Capella back in 1863. In America the irrepressible Henry Draper photographed the spectrum of Vega directly and obtained a good image of the absorption lines. It was becoming obvious to all what an important astronomical tool photography could be.

By 1878 the dry astronomical plate became a widespread medium. C.E. Bennett combined the salts with a neutral medium and produced faster dry plates for astrophotography. The next year (1879) George Eastman (of Eastman Kodak fame) built a machine that would coat glass plates with silver salts at a prodigious rate, effectively mass producing a quick standardized process of photography with faster emulsions ready for commercial use.

That same year the first quality photographs of Jupiter and Saturn were obtained, and in 1881 the first photographs of a comet (Tebbutt's comet) were made. The Orion Nebula had been photographed in 1880 by Henry Draper and revealed the power of the new photographic tools by giving the most detailed look at the gas and individual stars of the nebula yet produced. So good was the dry plate process that the first spectra of stars were photographed, again by Draper, enabling studies to be done at leisure.

Photography in astronomy took a quantum leap forward in 1899 when Julius Scheiner working at the Potsdam Observatory photographed the Andromeda Galaxy—or what he at the time referred to as a "spiral nebula." The exposure was 7.5 h long! By this time the observatory at Harvard was already completing its first photographic surveys of the sky, detecting stars down to the eighth magnitude.

The only problems with the early dry systems were the graininess of the images. With grain sizes within the solutions being almost 10 m there were limits to the details obtainable in any image. Notwithstanding this problem, the first survey of the starry sky implemented as an international project was begun in France in 1887 by Amedee Mouchez, the director of the Paris Observatory. The *Astrographic Catalogue* enabled stars to 11th magnitude to be recorded to provide an absolute standard reference for all astronomers. The dry plate process had come of age.

Coupled to the giant reflectors being built in the United States at the beginning of the nineteenth century, photography as a standard reference medium was becoming the pinnacle of astronomical recording.

Problems with graininess and faster speeds were eventually overcome by ongoing research until the whole photographic field was overtaken by CCD technology in the late twentieth century. Photography took a great leap forward with the development of large reflector telescopes in the early twentieth century and the production of purpose-built Schmidt cameras to obtain wide field images of the sky.

It was quickly realized that the color of an object was related to several astrophysical properties that were important, with some unexpected relationships being uncovered between various parts of an astronomical object, including differences in gas composition or dust content. Up to this point most astronomical photographs were monochrome, so to untangle these relationships a system of full color imaging had to be found.

Color photography is almost as old as the monochrome process itself. It was the physicist James Clark Maxwell that suggested three-color photography in the red, green and blue bands in 1855. However it was not obvious how this could be achieved until Thomas Sutton used a color separation process in 1861 that involved photographing a ribbon in monochrome through three colored filters—red, green and blue and combining the final image. However the process was not sensitive enough to red light, and Sutton's pioneering experiment was forgotten before it re-emerged in a reformatted technique used by the Lumiere brothers in 1903. Their process involved a three-filter system of red, green and blue dyed potato starch underlain with photosensitive silver halides to produce what they termed autochrome plates. This was a popular addition to ordinary photography up until the 1930 advent of the subtractive color film, but it was still not sensitive enough for astronomical use.

The inception of subtractive color film was a revolution in photography. The film layer was now a sandwich of three layers, each sensitive to red, green or blue areas of the spectrum and could then be processed, enlarged and turned into a photographic positive in a relatively easy chemical process. During the 1970s and 1980s color reversal film (slide film) and color negative film became a bit hit among amateur astronomers as the better technology, decreasing grain size and increasing sensitivity, coupled to driven telescopes and mounts, made astrophotography accessible. However for the professionals the three-filter system originating with James Clark Maxwell became the preferred method of astronomical photography, as it enabled them to produce standard images that revealed the subtleties of the underlying chemistry of astronomical subjects.

Once again Eastman Kodak was in the vanguard of these developments. By 1937 the company was offering astronomers photographic plates using a variety of photographic emulsions that were sensitive to over 20 spectral ranges. These were sold to observatories worldwide along with the recommendations on their uses in the in-house booklet, *Photographic Plates for use in Spectroscopy and Astronomy*. All of these emulsions were monochrome and were to be used with the correct filters to render as correct an image as possible. Thus the development of filters and astronomical emulsions went hand in hand.

Color photography at observatories then followed the familiar pattern of taking a monochrome image through three filters with red, green or blue sensitive emulsions underlying the filters. The negatives would then be contact copied together to make a color positive and then enlarged to give a final color image that could be studied. Such techniques attained their apogee under the hand of David Malin at the Anglo-Australian Observatory, where Malin's techniques of color stacking and unsharp masking brought hitherto unseen detail out of astronomical objects.

Reciprocity failure, blue and red sensitivity on emulsions, infrared absorption and a host of other potential problems were improved upon, and new photographic techniques such as the use of gas hypersensitization, cold cameras and the introduction of unsharp masking and image stacking made astronomical imaging an art in itself. Looking back on much of this development it is surprising to realize that all this would be swept away in a few short decades. Such wonders as T-grain technology in tech pan 2415 and others of its ilk seem now consigned to the dustbin of history, but in their day they were a great addition to the astrophotographer's toolbox and were a wonderful hand-down from the development of professional astronomical imaging and the evolution of chemical technology applied to photography.

CCD and DSLR Systems

The CCD, or charged coupled device, was invented at Bell Telephone laboratories by Willard Boyle and George Smith in 1969. Both were thinking of a chip that could store information for use in computers and were not thinking of imaging at that time, but by 1971 the image-processing power of the early devices became clear when they were demonstrated in national reconnaissance satellites in the mid-1970s. Due to their high efficiency at capturing photons of light they quickly became adopted by professional astronomy observatories worldwide and found application across a wide range of wavelengths from UV to infrared.

There are many CCD cameras on the market for novices and advanced amateur astronomers. Manufacturers such as SBIG (Santa Barbara Instruments Group), Starlight Express, Atik, Orion, Celestron and many others provide good quality cameras to suit most budgets. A typical example of a CCD camera is shown here in Fig. 1.1.

CCD cameras have a multitude of uses, from imaging an object to obtaining information and data through photometry, astrometry and spectroscopy. All require coupling to a computer, and most astronomers use laptops in the field or in purpose-built observatories to maximize the use of these excellent image capture technologies.

CCD cameras come in a bewildering array of chip sizes and shapes; one can have a square CCD chip containing 2,048 × 2,048 pixels or rectangular ones with pixel sizes varying from 1,500 × 800 and upward. Most professional CCD devices are enormous and have great sensitivity and resolution.

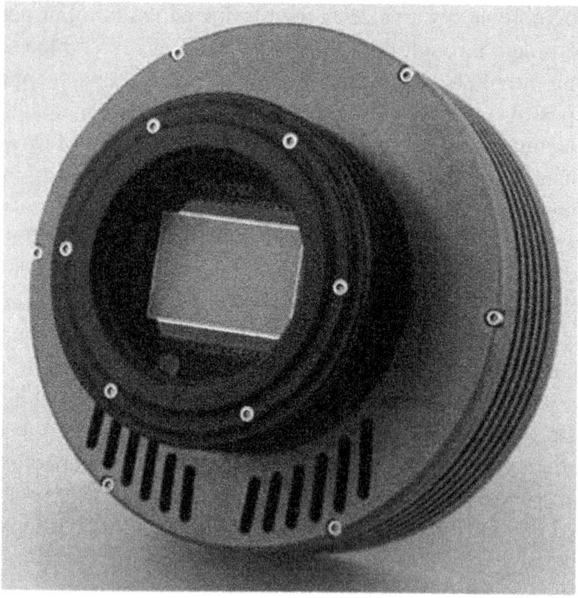

Fig. 1.1 Atik CCD camera

To obtain a full color image it is necessary to use a variety of filter systems with a typical monochrome CCD camera. Rendering the image through red green and blue (RGB) filters gives a natural color to most astronomical objects. However, as the sensitivity of CCD chips has increased so has the application to UV and infrared wavelengths, leading to CCD photography filters that cover these ranges. One would then have a set of filters covering UV between 320 and 400 nm (U), blue light between 400 and 500 nm (B), green or visible light at the center of the visible spectrum between 500 and 700 nm (V) red light and infrared light at increasing wavelengths from 800 to 950 nm (I), resulting in the combination of all these generally being referred to as UBVRI.

Ideally the observer who wishes to image the sky should be equipped with either a CCD camera or digital single lens reflex camera (DSLR). Choosing such equipment can involve a long process of comparison and getting advice from experts in the field, but such advice is well worth considering so as to avoid common mistakes. A DSLR is a versatile tool, and of course is very useful for photography outside that of astrophotography, whereas a CCD camera is not. There are many manufacturers, but the general consensus is to purchase either Nikon, Canon or Olympus cameras, as these manufacturers have a wide range of auxiliary equipment available such as lenses, T-mounts, adapters, focusing screens and filters. Although digital, these cameras follow a similar format as 35-mm film cameras in that their sizes, weights and controls are flexible, and they are easy to set and control once one has had some experience with them.

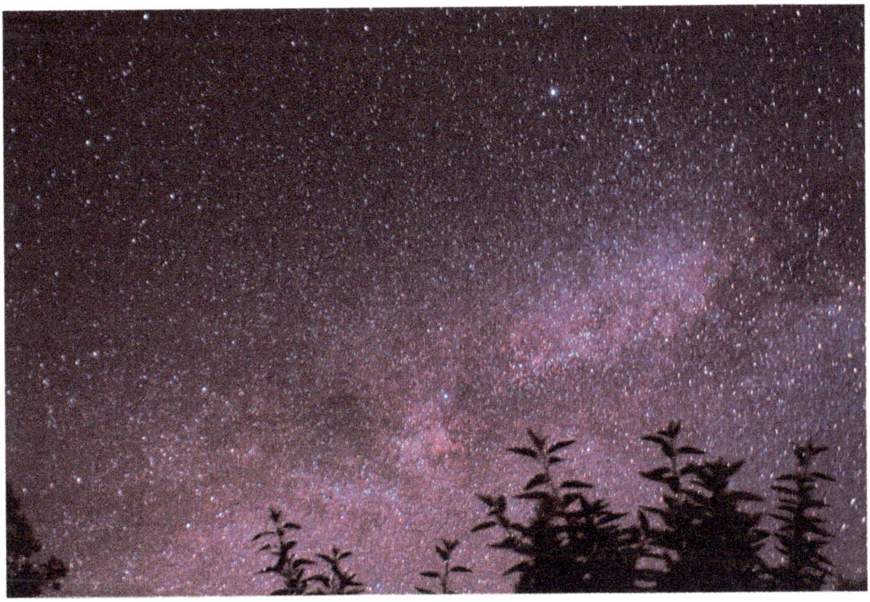

Fig. 1.2 Summer Milky Way in Hα

It is not our intention in this book to recommend any particular brand Excellent results can be had with all the above types; it is merely a choice of preference and cost. In addition to the camera, the observer will require a purchased cable release inimical to their camera system, to prevent shaking. If the observer intends to connect the camera to a telescope, then a T-mount and an adapter sleeve are also essential. Some suppliers provide readymade T-mounts fitted with a contrast or Baader or a Hα filter, depending on their prospective use. Conversely, several companies sell filters that fit over the front element of particular lenses for wide-field photography, which enables some spectacularly different pictures of objects such as the Milky Way, as can be seen in Fig. 1.2, where Andy Burns of the Wilshire Astronomical Society took an Hα shot of our galaxy in summer.

There are several advantages of the DSLR camera over that of the CCD, such as the quick response time of displaying the image, the amount of technical control over the camera by an experienced observer, the ability to take wide-field shots through a variety of lenses and the ability to vary the ISO ratings is an advantage. The lower thermal noise in a short exposure and the storage on an SD or CF card for later use make the laptop superfluous at the telescope and provide a good option in favor of a DSLR. However, on balance the sensitivity and range of a CCD camera and its cooled chip with small thermal noise in long exposures make the CCD camera a preferred tool among astrophotographers.

Filters for Astronomy

To provide these filters a specialist industry has grown up to serve both professional and amateur astronomers alike. The technology has enabled filters to be built that can target particular areas of the spectrum and look at narrow bandwidths or block out areas of the spectrum where extraneous external light could interfere with the observations. However these are not the only filters available, and many visual astronomers, especially in the amateur market, use colored filters to enhance their view of planets and some deep sky objects.

Again, it is not the intention of the author to recommend any particular filter or system of filters, as there are so many commercially available. As the reader will discover as one reads this book, filters can be object specific or generalized for photography. There are filters for the Moon, filters for the Sun and filters for the planets that can all be used for visual or photographic work. Broadband and narrowband filters for observation and imaging are ubiquitous in today's astronomy marketplace.

It is up to the individual to determine their field of study and to decide which, if any, filters to opt for. Visual observing, apart from observing the Sun, needs very little from a general viewpoint. Amateur astronomers who contribute greatly to fields of study such as planetary work regularly use filtration systems based on colored filters that have come to the fore for photographic use. Other astronomers will opt for narrow and broadband filters to perform imaging and use techniques that give so many today observatory-quality images at a fraction of the cost and time of obtaining images in the past from photography alone.

This author hopes that the following sections will enable readers to make a choice, or at the very least to be informed of the filter systems available today for their chosen craft.

Further Reading

Pedro Re (2009) History of astrophotography timeline. http://www.astrosurf.com/re/history_astrophotography_timeline.pdf

Royal Astronomical Society (2012) The history of astronomical imaging (specialist discussion meeting, January 2012)

Trow A (2006) Photography was Hijacked by astronomers. Astronomy Now Magazine (March 2006)

Chapter 2

Filters and General Equipment for Astronomical Observing

If you are an avid visual astronomer, it may come as little surprise to find that colored filters make a world of difference to your observing clarity, and their use brings out detail from the subtle shadings found on Solar System objects.

On occasion, a telescope manufacturer will include a set of colored filters along with the telescope you have purchased, although many amateur astronomers tend to ignore their use, and the filters just gather dust as the telescope becomes well used. If the filters are of good quality then they can be used to enhance the view of some objects quite easily after a little patience and trial.

Colored filters with a wide bandpass are a very useful tool and should not be abandoned simply because one has not yet used them or has had a little practice. They can make a great difference between seeing or recording an object or missing it completely in the sky background. Many planetary observers rely on filters commercially available, as they report that they really do make a difference in seeing faint details and also assist with reducing glare, as many of the planets and especially the Moon are very bright and suffer from the enormous contrast between their sunlit surface and the dark background sky.

Filters for Visual Observation

Astronomical filters for visual observing are a specialized piece of the astronomer's armory. They work by blocking a specific part of the color spectrum, usually an "opposite color," which then leaves the remaining wavelengths a little more open to view. The color of the filter lets through wavelengths that correspond to that color while darkening or providing more contrast to wavelengths outside of the filter

M. Griffiths, *Choosing and Using Astronomical Filters*, The Patrick Moore Practical Astronomy Series, DOI 10.1007/978-1-4939-1044-1_2,
© Springer Science+Business Media New York 2015

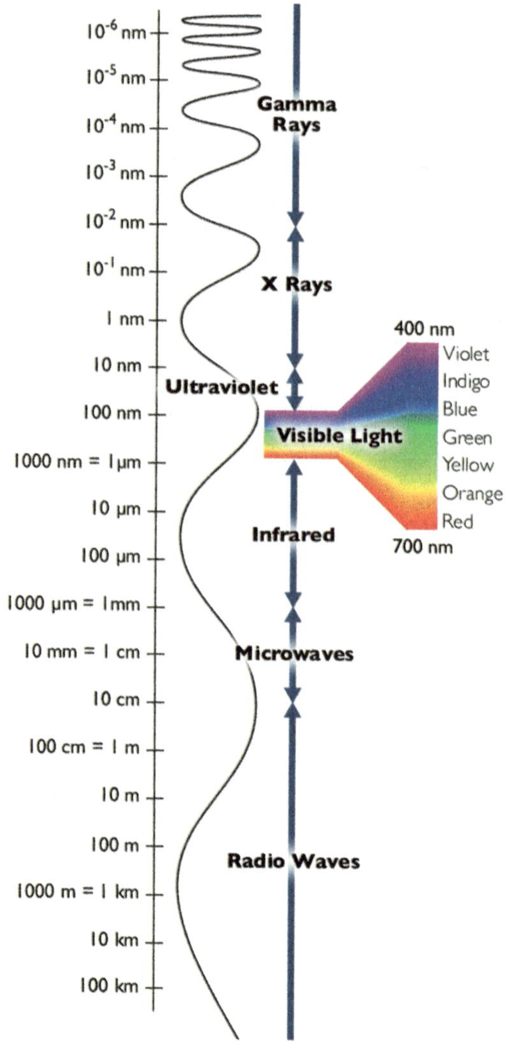

Fig. 2.1 EM spectrum

range or blocking them altogether. The visual element of the electromagnetic spectrum can be seen from the accompanying Fig. 2.1.

As can be seen the visual part of the spectrum lies in a narrow range between 400 and 700 nanometers (nm), with blue at the shorter wavelengths and red at the longer. Using a red filter, for example, with a bandpass longer than 600 nm will block wavelengths of light short of 600 nm and will render such shorter wavelengths as black or at least darker and give increased contrast to details in an astronomical object. One must also remember that many of these details are subtle, and use of filters may enhance some features but render others almost invisible. It's a matter of trial and error.

The colored filters that will be mentioned shortly are best used on Solar System objects via visual observing. They do not necessarily work on CCD imaging systems outside of the BVR (RGB) range, and neither do they work for DSLR cameras unless used with the planets and augmented with skills in image processing. For deep sky work they also have their limitations until you get into the range of narrowband filters where the specific bandpass enables the photographer to build an image based on true color and wavelength rather than relying on color or BVR, as one would in terrestrial photography. Please note that the filters detailed below are largely for visual work only. We shall deal with the use of filters in photography in another chapter.

Wratten Filters for Lunar and Planetary Observing

Although there are several astronomical suppliers that provide these filters with generic names such as Meade, Agena or Orion, all such filters are evident by their color and are usually marked with particular numbers known as Wratten numbers, which allow the observer to choose which parts of the EM spectrum they are going to enhance in order to make planetary and lunar definition and contrast easier to discern through the eyepiece.

The Wratten system was developed in Britain in the early twentieth century by Fredrick Wratten and Kenneth Mees, who founded a company in 1906 that produced gelatin solutions for photography. Mees then developed gelatin filters dyed with tartrazine to produce a yellow filter, but soon developed other colors and a panchromatic process of photography. In 1912 they sold the company to Kodak at Harrow in England, and Mees moved to New York to found the Eastman Kodak laboratories there. In honor of his partner and mentor, Kenneth Mees named the burgeoning number of colored filters "Wratten" and introduced the complex numbering system that is still in use today.

Not all the Wratten filters are suitable for astronomical use, but the main colors are still widely used in visual astronomy and terrestrial photography work and are detailed below.

These colored filters are known as broadband or "longpass" in that they allow a large variety of wavelengths through but block wavelengths above or below a certain range in the EM spectrum. As the spectrum in visible light lies between 390 and 700 nm, with the blue wavelengths being the shortest (~400 nm) and the red being the longest (~700 nm), then anything with a wavelength range above or below a particular filter will be blocked and increased contrast in compensating colors will be noticed.

Most astronomical suppliers sell complete sets of filters for Solar System observing, and naturally such sets are known as lunar and planetary filters.

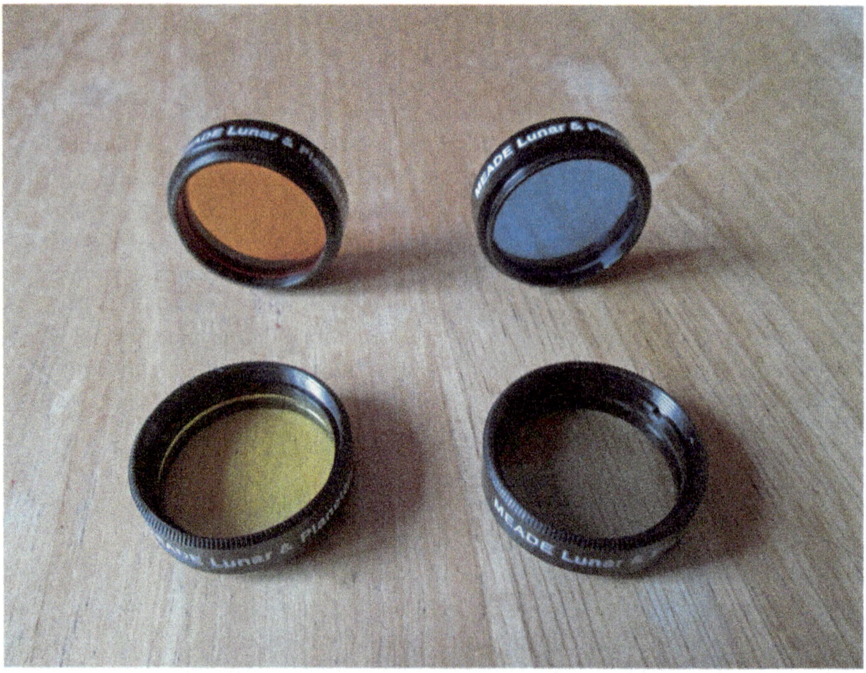

Fig. 2.2 Meade lunar and planetary filters

They have a range from red to blue across the spectrum and cover the broad bandwidths associated with such colors in addition to covering some of the wavelengths of the Wratten filters that are discussed below. A typical set will include a neutral density filter for lunar observing and a No. 25 red, No.12 yellow and No. 80A blue for as full coverage as possible. A Meade filter set can be seen in Fig. 2.2, although each manufacturer generally follows the same colour set for such work.

In the following section you should note that the Wratten number comes first followed by its color. However, we have grouped the filters under their color rather than put them in number order, as color is their most obvious feature when using them. We will use the spectral sequence from long to short wavelengths as the basis of the description, so we shall follow the standard ROYGBIV spectra that you probably encountered in school. There is also included some brief advice on the usefulness of the filter in visual astronomy before we move on to exploring the use of such filters in greater detail. All of these filters are available to purchase in 37.1 mm (1.25″) or 50 mm (2″) fittings for the observer's eyepiece range and are commonly available from astronomical suppliers (Fig. 2.3).

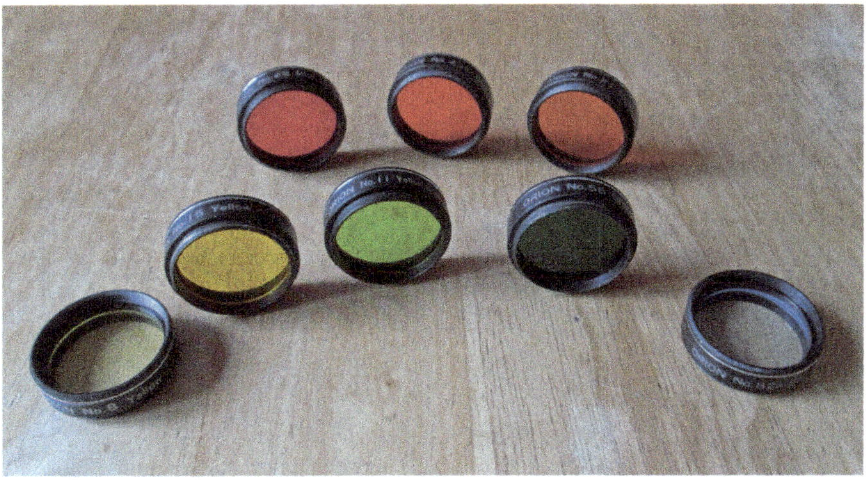

Fig. 2.3 Wratten colored filters

No. 25A Red

The No. 25A filter reduces blue and green wavelengths, which when used on planets such as Jupiter or Saturn result in well-defined contrast between some cloud formations and the lighter surface features of these gas giants. However, it needs to be used judiciously, as the light transmission is only 15 % and requires quite a large aperture, at least 150 mm+ for visual observation. It is also used to enhance infrared photography on a terrestrial scale, but for astronomical purposes it blocks light shorter than 580 nm wavelength.

No. 23A Light Red

This is a good filter for use on Mars, Jupiter and Saturn and may be useful for daylight observations of Venus as it has a 25 % light transmission through this rather dark filter, and, as it is an "opposite" color to blue, it darkens the sky very effectively in daylight. Some astronomers report that it also works well on Mercury, but this author would not recommend viewing this planet in general during daylight due to its proximity to the Sun, unless one is confident of their observing ability and equipped with a GOTO system. This filter blocks wavelengths of light shorter than 550 nm in the visible EM spectrum.

No. 21 Orange

This orange filter reduces the transmission of blue and green wavelengths and increases contrast between red, yellow and orange areas on planets such as Jupiter, Saturn and Mars. It brings out the glories of the Great Red Spot on Jupiter very well under conditions of good seeing on a modestly sized 'scope (150-mm aperture) with a median magnification. It also blocks glare and provides a lesser contrast between a bright planet and the background of space, as this filter only transmits about 50 % of the light and blocks wavelengths short of 530 nm.

No. 8 Light Yellow

This filter can be used for enhancing details in red and orange features in the belts of Jupiter. It is also useful in increasing the contrast on the surface of Mars, and can under good sky conditions aid the visual resolution of detail on Uranus and Neptune in telescopes of 250 mm of aperture or larger. The No. 8 is also quite useful in cutting down glare from the Moon during visual sweeps of this object and works much better than the "moon filters" usually included with some cheaper telescopes on the market. The No. 8 filter allows 80 % light transmission and blocks light short of 465 nm.

No. 12 Yellow

This filter works on the principle of opposites described above, blocking the light in the blue and green region and making the red and orange features on Jupiter and Saturn stand out more clearly. Deeper in color than the No. 8 filter, it is the filter most astronomers recommend for visual work on the gas giants. It has a 70 % light transmission and cancels some of the contrasting glare on Jupiter when seen against a dark background sky. It blocks visible wavelengths short of 500 nm.

No. 15 Deep Yellow

This filter can be used again to bring out Martian surface features and the polar caps in addition to bringing out detail in the red areas of Jupiter and Saturn. Some astronomers also have reported some success using this filter to see low-contrast detail on Venus. Other astronomers, including the author, have used this filter on Venus during the day to add more contrast to the image. This filter is particularly useful, as Venus is a very bright object, and the filter has a 65 % light transmission with a longpass blocking light short of 500 nm and can considerably reduce the glare of this very bright planet.

No. 11 Yellow–Green

This darker filter is a good choice to enable the observer to directly see surface details on Jupiter and Saturn. It can also be useful on Mars if you are using a large aperture telescope in the 250 mm+ range and the seeing is steady as it darkens the surface features and makes areas such as Acidalia and Syrtis Major stand out well. The No. 11 filter allows 75 % transmission of light and can be used to darken features on the Moon, too. However, it is more of a color correction filter rather than a longpass filter and will allow all visual wavelengths through.

No. 56 Light Green

This filter has been used by the author for observing the ice caps of Mars during its close encounter in 2003 and found that despite the low altitude of Mars from the UK during that apparition the filter worked well in bringing out these features and even hinted at other rocky features on the planet's surface during periods of clear seeing. True, the orange filter above did work better in rendering color and detail on the Red Planet, but the contrast with this filter was quite good. This will allow most wavelengths through but does have some concentration around a wavelength of 500 nm.

With its 50 % light transmission this filter is a favorite of lunar observers, as it increases the contrast while reducing the glare. It is also a filter that is well tuned to the wavelengths of the human eye, and the greenish cast can almost be ignored during visual observation. This is another color correction filter with all wavelengths equally affected. The effect can be seen on the first quarter Moon in Fig. 2.4, photographed here in ordinary white light and then through the Wratten No. 56 filter. The glare is reduced and features are easier to make out in general, though it can be difficult to get used to seeing a green-colored Moon. Perhaps the ancients had something when they thought it was made of cheese…

No. 58 Green

This filter blocks red and blue wavelengths of light, and many observers find that it slightly increases contrast on the lighter parts of the surface of Jupiter. This author has also used it on Venus, where it does add to the contrast and reduces glare, but it must be admitted that it is not easy to visualize any detail in the clouds. However, because of its 25 % light transmission, this filter requires a larger aperture telescope—probably above 200 mm is best. It is also a color correction filter rather than a longpass.

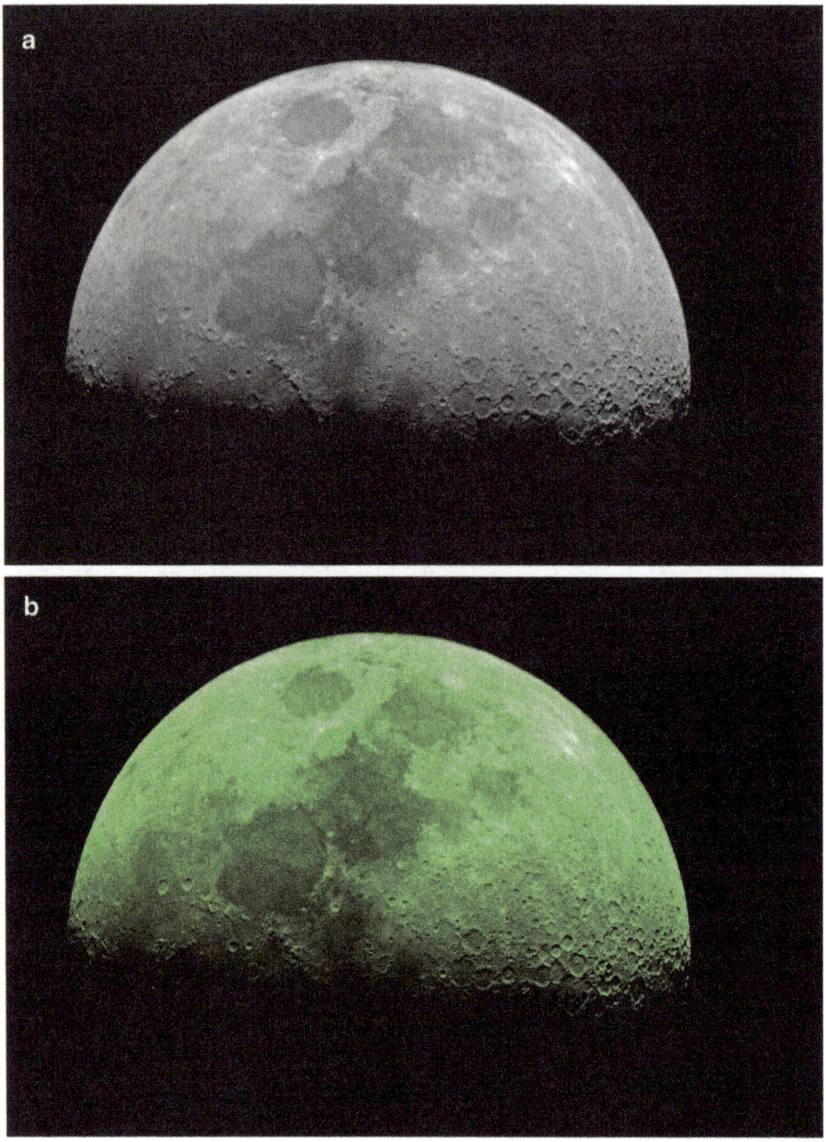

Fig. 2.4 (**a**) First quarter Moon, and (**b**) No. 56 filtered Moon

No. 82A Light Blue

This is almost a multipurpose filter, as it does enhance some features on Jupiter, Mars and Saturn and also works quite well in enhancing some features on the Moon. It is commonly referred to as a "warming" filter that increases the color temperature slightly and allows the red wavelengths through. With a light transmission of 75 % it can be used on any aperture telescope and can even make some difference to deep sky objects such as M42 and M8, though the effects can be quite subtle.

No. 80A Blue

Although this is quite a dark filter, it is as versatile as the No. 82A in that it enhances red features on planets such as Jupiter, Saturn and Mars. It is also good for lunar observation, as it reduces the glare and provides good contrast for some features such as ejecta blankets, ray systems and lava fronts. Some astronomers report success in its use on binary star systems with red components such as Antares and α Herculis as the contrast enables the observer to split the two components well. The No. 80A filter has a 30 % light transmission and so should be used with a larger aperture if possible, though this caveat does not apply to lunar observations. It also acts as a color conversion filter, enhancing wavelengths around 500 nm.

No. 38A Dark Blue

Again, a good filter to use on a planet such as Jupiter because it blocks red and orange wavelengths in such features as the belts and in the Great Red Spot. Some astronomers report that it also adds contrast to Martian surface phenomena, such as dust storms, and makes a better contrast for observing the rings of Saturn, though this author has not observed a dust storm and can see very little effect on the rings of Saturn. However, some astronomers report that using this filter increases the contrast on Venus, leading to the visual observation of some dusky cloud features. This filter again needs a fairly wide aperture, above 150 mm to operate effectively, as it has about a 15 % light transmission. It absorbs red, green and UV light and is commonly referred to as a minus green plus blue filter.

No. 47 Violet

A very dark filter that strongly blocks the red, yellow, and green wavelengths. It is highly recommended for Venus observation due to its low light transmission of about 5 %, providing great contrast and enhancing cloud features. It can be used on

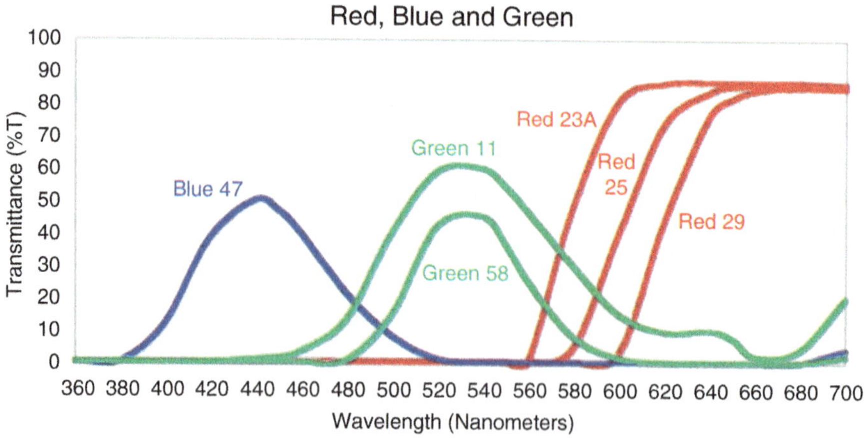

Fig. 2.5 Wratten filter bandwidths (*Source*: Analog Photography Users Group (APUG) at http://www.apug.org/forums/forum45/60828-tiffen-filter-wratten-equivilents.html)

the Moon to decrease the glare when observing features at a 10–12-day-old phase, and some features in the Schroeter Valley and Aristarchus Crater were better seen due to the lack of glare. Recommended for the Moon if you have a large aperture! This is another color separation filter that enhances the blue or shorter wavelengths of the spectrum.

The above filters enhance the astronomer's observations of the planets and our lunar neighbour very well and are highly recommended. Figure 2.5 gives the typical bandwidths of the Wratten filters considered above.

Additional filters that also are helpful in visual observing are the polarizing filter and the neutral density filter. Both are longpass filters that usually transmit all wavelengths of light but can cut down on glare and contrast.

Neutral Density Filters

A neutral density (ND) filter transmits light uniformly across the entire visible spectrum and is an excellent filter to use to reduce glare in objects such as the Moon and planets, especially the Moon. Due to its bright glare many lunar and planetary astronomers keep a permanent ND filter on their favorite eyepiece and add on other filters as necessary. The neutral density filter is quite handy if you're are doing a public observing session where you can switch from a first quarter Moon to deep sky objects, as its reduction in glare assists in maintaining your night vision and doesn't result in customers squinting and half blinded as they stumble away from the bright object in the eyepiece! These filters come in a variety of densities that apply different contrast to the image based upon the amount of light transmission each ND filter allows. Commonly they come in numbers such as 50, 25 and 13 that signify the amount of light they transmit—50, 25 or 13 %, respectively.

Polarizing Filters

Although it does not work at any particular wavelength, the polarizing filter allows light of any wavelength through but blocks those with random scattering patterns, allowing only light waves in a flat "plane" through, which has the effect of increasing the contrast, reducing glare and slightly enhancing the saturation of color in an object. Such filters are very useful on bright objects such as the Moon and planets, and we shall examine their effects in due course.

Why Use Filters for Visual Observing?

The above information should provide the visual observer with a little guidance on using filters to see subtle details on some Solar System objects. Not to be forgotten is the fact that occasionally you can combine filters to produce a better visual view, though be careful that you do not end up with a set of filters that will give you a neutral density! Combining some red green and blue filters will do this—and in doing so the observer may as well remove the filters altogether, as all they have done is cut down on the light transmission through the telescope! This may be fine for viewing the Moon, but for fainter Solar System objects, maximum light transmission and enhance contrast through using a filter is preferable.

Some observers may scorn the use of Wratten filters and can claim to see enough detail in objects without their use. To this argument the author can only say that any optical aid is useful if it achieves the goal of the observer. The experience of many seasoned observers reveals that using colored filters can overcome such problems as image deterioration due to atmospheric effects and allows separation of colors in a planetary atmosphere that may not be attainable by the human eye alone while staring at the glare of a bright Solar System object such as Jupiter or Venus. They provide a welcome contrast between areas of different colors and provide a sharper less "glare-filled" view of a bright object.

Astronomers must remember that the human eye has not evolved to cope well in the dark, and our color perception departs noticeably from the norm once we become dark-adapted. Subtle colors such as reds, browns and greens disappear at night as the human eye becomes adapted to the blue range of the visual spectrum, a phenomenon known as the Purkinje effect after its discoverer the Czech anatomist Jan Purkinje. He noticed that under low illumination the human eye loses the detection of wavelengths longer than 555 nm as the chemical opsins and rhodopsin in our eyes have a peak absorption of around 500 nm. Observers end up with essentially black and white vision, so the addition of a colored filter with a wide wavelength that enables one to make the most of the "missing" wavelengths is going to be an advantage in any observing situation.

In addition to the above, the observer has to contend with problems related to the atmosphere, through which we see any astronomical object. Atmospheric scattering is responsible for an almost luminous veil through which we visually perceive an object and is due to the nature of the particles in the air. Atmospheric scattering is inversely proportional to the fourth power of the wavelength of light, so if we are viewing an object that is primarily blue in color with a peak wavelength of 480 nm, then the light is scattered more than a red object at 750 nm. This is why our sky appears blue during the day. Remember that at night we lose the red end of the EM spectrum due to the Purkinje effect! A filter can add enough contrast to partially overcome the effects of scattering.

Observers are also affected by a phenomenon known as prismatic dispersion, which is most evident when a star or planet is seen near the horizon. It results from refraction in the atmosphere being less for the longer wavelengths, whereas the red appears nearer the horizon and violet toward the zenith. This is why most astronomers prefer to observe an object when it is near or at culmination so that this effect is lessened. Use of red or blue filters on an ascending object may make the difference between seeing details such as the Great Red Spot, for example.

Finally we have to counter the effects of irradiation, which is manifested by areas of unequal brightness, resulting in the brighter zones encroaching on the fainter areas within a planetary atmosphere. This effect is primarily psychological, as it is the interpretation of the brain that evens the illumination out, but results in decreased contrast and a lack of detail between areas of different brightness. Use of a red or orange filter will even out the illumination and allow the eye/brain coordination to distinguish differences.

In the case of bright objects the reduction in light transmission is an advantage. The difference in contrast between the belts and zones on an object such as Jupiter can be so great that the human eye and brain just smear the whole, and it can be difficult to discern details without a filtered system. This is due to the differences in intensity of reflected light from light/dark zones in such planetary systems. To illustrate this let us examine the difference in intensity as viewed through a telescope.

Contrast in any system can be measured using the formula:

$$C = \left(b2 - b1\right) \div b2$$

where C is the contrast and b1 and b2 are different areas of brightness on the surface of a planet. Bright areas on Jupiter have an intensity of 6 Lumens m^{-2}, and the intensity of the darker zones have an intensity of 3 Lumens m^{-2}. This would give:

$$\left(6 - 3\right) \div 6 = 0.5$$

a contrast 50 % lower in the darker zones than in the brighter zones. A filter will even out the contrast by permitting wavelengths representative of the redder or darker zones through while diminishing the contrast on the brighter zones. Surely a filter that would aid in the perception of subtle features is going to be a bonus to any observer?

Therefore the use of filters, despite their decrease in light transmission, is very useful in visual astronomy. The use of filters assists primarily in reducing contrast initially, and though the reduction in light transmission is generally not favored in astronomy, this is one area in which this general rule need not apply.

It takes time and personal training to use a filter effectively and to discern different details in the atmosphere of a planet such as Saturn or Jupiter, let alone fine details on the Martian surface. Nevertheless, filters offer a very convenient and relatively cheap way of seeing these details, so the application of a little effort is well worthwhile.

Observing Hints: Dark Adaptation, Telescopes and Eyepieces

The above filter groups must be used in ways in which their maximum utility can be gained. The most important considerations when using a filter set are not only their effect on the wavelengths of light passing through any optical system but also the types and aperture of the telescope in use, the magnification possible with such an instrument plus the amount of dark adaptation that the observer can achieve. Secondary considerations are the amount of pollution and dust in the atmosphere, atmospheric transparency, the air humidity and the steadiness of the atmosphere at different times of the year.

Dark Adaptation

Most people underestimate the power of the naked eye and how to use it properly to gain the most from astronomy. Telescopes, cameras and CCDs are designed to get the best from any observing or imaging, but the human eye is the centerpiece of any attempt to see the sky, and many are surprised at what the eye can discern. On a dark night, the iris, the diaphragm controlling the amount of light entering the eye, is fully open, and measures on average 6 mm in width. This maximum aperture is most desirable to astronomers, so when you go out into the darkness of an evening, take a little time to allow the iris to open fully, a technique known as dark adaption. This process can take about 5–10 min, but to be fully dark adapted takes up to 30 min.

During this period of time, do not go near any light sources, switch on your flashlight or stare at car headlights, otherwise your adaption will be ruined and you will have to start again. Many observers supplement this process by closing their eyes, thus hoping to speed up the adaptive procedure. A fully dark-adapted eye is obtaining the maximum input from the surroundings and can discriminate between subtle areas of light and shade on most astronomical objects.

Once this has been accomplished, you can begin observing, as your eyes are now a little bit more sensitive to light than they ordinarily are during the day. However, some astronomers have noted that faint objects seem to be a little brighter if they are seen "out of the corner of the eye," as it were, and this phenomena is another important tool of the observer. The process of seeing objects in greater detail simply by not looking directly at them is called *averted vision* and is a bonus when looking for faint objects such as galaxies or planetary nebulae or even faint details on planetary surfaces. This phenomenon arises due to the way that the eye is constructed. At the rear of the eye is a light-sensitive membrane called the retina. The retina is fabricated from two sets of cells, commonly called "rods" and "cones." The cones lie directly behind the pupil when the iris is fully open and the light sensitive rods lie off to the sides of this aperture. So when not looking directly at an object the more sensitive rods are able to pick up the stray light that the cones are missing, thus making averted vision a good habit to get into as far as astronomers are concerned.

Thus when not looking directly at some object, the more sensitive cones are able to pick up the stray light that the rods are missing, thus making averted vision a good habit to get into as far as we astronomers are concerned. To enable you to maintain your night vision, it is best to examine any star charts or atlas by means of a flashlight with a red beam. This red light will not interfere with a dark-adapted eye and is comfortable and easy to see such charts by Equatorially mounted Reflector telescope.

When observing, make sure that your comfort is the paramount consideration, so always dress warmly, have a hot drink handy and take a break every hour or so if you intend to observe all night. Additionally, if you can stand on a raised board while observing, then the heat of your body will not be sapped through your feet, leaving you cold and miserable. This is simple common sense, you may well say, yet many observing sessions have been ruined by the lack of such preparation.

To enhance your appreciation for nature's deep sky wonders, and to bring out the best in your optics, it is imperative to find a dark sky site. What is meant by this term is a site where no street light or the pervasive orange glow of streetlights can be seen, and the sky remains dark right down to the horizon without interruption from any light sources through a 360° circle. Granted such sites may have to be found well outside your town or area, especially for urban astronomers, but it is worth the effort to go out of your way to obtain fine observational results. The International Dark Sky association has designated dark sky parks and reserves in many areas of the globe, and it would be useful to locate the closest one to you. If this site is too far away, local astronomical societies have information on local areas that are fit for observation.

To see deep sky objects and faint details at their best, it is best to avoid times of the month when the Moon is shining brightly. Although the Moon is a lovely romantic object shimmering with a silvery light, looking wonderful in a cloudless sky, it is less than romantic to astronomers and astrophotographers interested in digging out remote or obscure clusters, nebulae and galaxies. However, for planetary observing the phase of the Moon is an irrelevance; many observers report that a bright moonlit night is ideal for planetary observing, as the darker sky background is washed out, resulting in less contrast between the bright planet and the deep sky. However, only observe when the air is fairly clear and the atmosphere, or seeing conditions, remain quite steady.

Telescopes

If you own a small telescope, then ensure that it is properly mounted and the mount is as stable and vibration-free as possible. A steady mount cannot be overestimated, as any vibration is going to take time to damp down and render a good clear image. Many small telescopes have alt-azimuth mounts that need constant adjustment to keep up with the rotation of Earth and thus have a lot of unstable vibration to contend with throughout the observing session. Tightening up any bolts and nuts on the axes and making the mount stable by hanging weights from the center of the leg supports can have positive results, as can having good slow motion controls. Better stability can be achieved with an equatorially mounted telescope, even if undriven, as the slow motion controls ensure use of only the right ascension axis while a driven mount that will follow the target is much to be preferred. In addition, the axes on equatorial mounts can usually be locked down, and slow motion controls are a little finer with less vibration.

The market is flooded with good-looking telescopes that perform extremely poorly and are purchased by the unwary with an interest in astronomy but no knowledge of optics. Ask anyone in a local astronomy group for the best advice on what to look for when purchasing a telescope. Good optics are an essential when using filters, as bad quality optics will destroy the fine details that a filter is meant to bring out.

When buying a small telescope, do not believe the claims of the manufacturer that this instrument will magnify up to 400 or 500 times. Such claims are almost always fraudulent, as at such magnifications only a blur will be observed through the eyepiece. As a general rule, a telescope is performing at its optimum when it has a magnification of 25× per 25 mm of aperture. Therefore, if you have a 100-mm telescope, the maximum it should be able to magnify, with resultant clear detail, is 100×. Following this advice will forestall any frustration you will ultimately experience if you have the misfortune to be sucked in by the advertiser's gimmicks. This is just a general rule, however, and depends on the seeing conditions and the quality of the telescope, so experiment with differing magnifications as you see fit.

Another key element in any visual system is the quality and clarity of the optics in the telescope and eyepieces. Image definition depends on sharp focus, resolving power and the contrast that can result from different levels of magnification, so keeping your telescope in top condition is a must. Clean optics with good eye relief to the eyepieces is an essential part of astronomical observing. With the plethora of telescopes flooding the market today amateurs can easily maximize the size of their telescope for a small financial investment. However, this does not always work to the advantage of the observer. Do not sacrifice quality for quantity. A large telescope with a 200-mm aperture may sound ideal to most observers, but the increased resolution may be sacrificed to the increased glare and lack of contrast unless a filter is used on a bright object.

In many cases the small and faint details that are rendered visible by a filter, especially on such planets as Mars, are dependent on the resolving power of the telescope, and naturally the larger the telescope, the greater its resolving power. Tiny details in the sub-arc second range will only be evident under high magnification with an aperture of the correct resolving power. To calculate the resolving power of any telescope the formula is:

$$\alpha = 0.25(D/\lambda)$$

Where α is the resolution in arc seconds ("), λ is the wavelength of light in microns and D is the diameter of the telescope in meters. Using such a formula we can obtain the resolution of any telescope we wish with a simple calculation. If λ is the middle of the optical spectrum at 500 nm, in microns it will be 0.5 and the diameter D of a telescope is 100 mm or 0.1 m, then dividing λ by D we arrive at 5, which now must be multiplied by 0.25, which gives us an angular resolution of 1.25" or 1.25 arc seconds. A telescope with an aperture of 300 mm will give a resolution of 0.4 arc seconds, so filter use with a larger aperture not only allows the observer to clarify and enhance the detail but also see increased detail at higher resolution than a smaller telescope.

Although it is impossible to recommend a specific type of telescope, as refractors, reflectors and catadioptric all have their uses, it is obvious that size is an important consideration. Nevertheless, a small refractor will generally give good clear views if the optics are of good quality, and for their size and portability they are very hard to beat. Others will rely on reflector telescopes, which may have very large apertures in comparison to the 70–100 mm of small refractors. Obviously the aperture and optical quality of such telescopes ranges widely, but no matter what the system in use, the addition of filters when observing the Solar System are a bonus in terms of the detail, clarity of the image and the ability to discern small features on a planetary disc. Let us examine each type in turn.

Refractors

A good quality refractor in the 100–120-mm range is a very versatile instrument and will provide a good platform for observing most deep sky objects. Refractors of this size give much better images than reflectors and even some Maksutov-Cassegrains or catadioptrics, as the light path is unrestricted and provides a ready platform for an SLR camera or a CCD device.

A point to note with refractors is that their focal ratio (f) can vary between types. Some of the short focus refractors such as the Skywatcher Equinox ED-Pro are superb instruments made with the best quality optics, but some of their 80-mm scopes are a fast f6 ratio, which means that the field of view through the instrument is quite large, spanning more than 2° with a low power eyepiece. Although this is not necessarily a consideration for most observers, should you wish to image or visually observe planetary nebula then this large field of view may lead to some confusion in identifying objects that are only a few arc seconds in size. A refractor with a longer focal length (f10, for example) may be a better requisite for such observations or imaging.

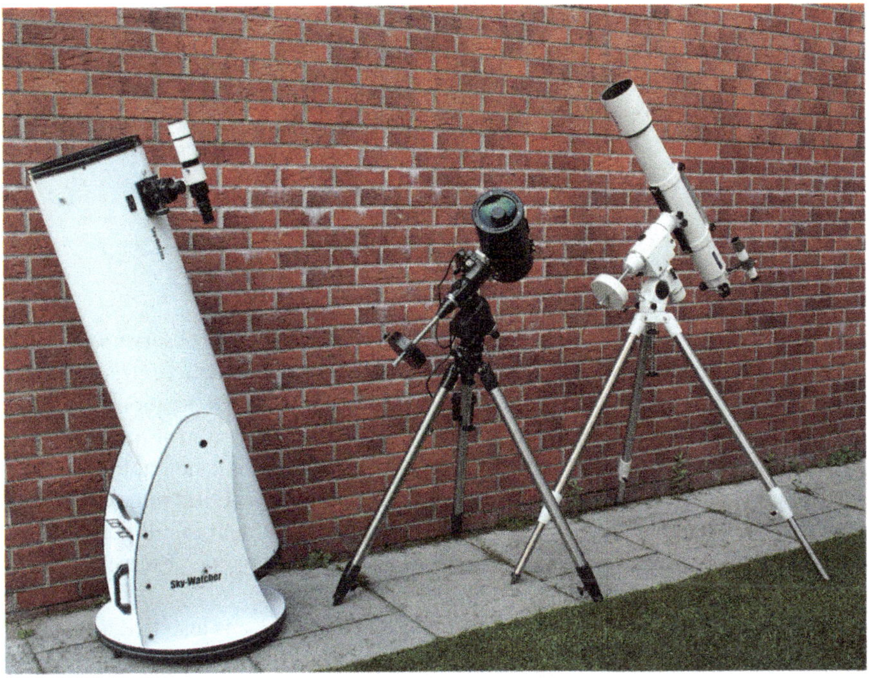

Fig. 2.6 Types of telescopes

The drawback of such an instrument is its limited aperture. A good refractor is a versatile instrument, but the difficulty of making quality objective lenses in sizes larger than 150 mm for commercial sale has always provided the amateur with a financial problem. To see really faint and indistinct objects, a reflector rather than a refractor may become the instrument of choice, as reflectors are very durable, portable, despite their larger size, and the sheer size of the aperture and light grasp plays into the hands of those looking for fainter objects or more detail in the brighter ones. Figure 2.6 shows a typical selection of telescopes available from most suppliers.

In most amateurs' experience, the quality of the optical system is paramount. Refractor telescopes using fluorite or Schott glass lenses are among the best available, as they reduce chromatic aberration and render very clear, well defined color images that can take a relatively high power. The only drawback with such systems is that they generally tend to be relatively limited in size outside of major observatories, and some telescopes, such as the wonderful Televue and Takahashi systems, command high costs, though the resolution, clarity and quality of the resultant image visible through such systems is second to none. A good refractor telescope is a prized possession.

Aperture, stability and quality ensure that refractor systems can be coupled to the whole range of Wratten filters so that the observer will greatly benefit from their use when observing the Moon and planets. DSLR and CCD imaging can be performed quickly and relatively easily with a filter wheel and camera placed accessibly at the main focal point of the telescope, though the observer needs to beware of the tripod legs on some mounts when imaging objects overhead!

Reflectors

Reflectors have become the main telescope of choice for most observers, as large apertures can be purchased for a fraction of the cost of a top-quality refractor. Most reflectors are built according to the Newtonian design, where a parabolic mirror at the base of the telescope reflects light back up the tube to a mirror angled at 45° (a flat) and then out through the side of the tube at a comfortable height for viewing. For the price of a good quality refractor you can buy a 250 or 300 mm reflector or larger on a Dobsonian mount. Despite their size, the instrument is still portable and can be transported relatively easily, though it is incumbent on the observer to check that the optics remain collimated when you set up.

A difference between refractors and reflectors in practice is also that a refractor is "ready to go" virtually right after set up, whereas a reflector may take some time to cool down to the external temperature before it can obtain fine images. Tube currents play a pervasive role in visual astronomy, and it is best to let a reflector settle before attempting to view any fainter objects on a target list. In addition, many of the Newtonian reflectors available commercially are not built for photography but just for visual observing. Although this is not a concern for most observers, it is something to be taken into account if one uses a large reflector. A Dobsonian mount negates the ability to track the sky effectively and so makes astrophotography a moot issue. A reflector on an equatorial mount overcomes such problems, but then the size and weight of the complete system begin to work against it as an easily portable object.

The sheer aperture of some reflectors make the use of filters a priority when viewing Solar System objects, as the large light grasp provides a very bright image that can suffer from glare and lack of contrast. On objects such as Venus, Jupiter, Saturn and especially the Moon, the low light transmission of the Wratten filters Nos. 25, 38, 80A and 56 enable detail to be seen without any loss of resolution.

Catadioptrics

These are almost a compromise optical system between reflectors and refractors in that they use both mirrors and lenses to achieve focus. Most amateurs will be conversant with the typical setup of such telescopes, such as the Schmidt-Cassegrain, wherein a corrector lens at the front of the 'scope adjusts the light path to fall on a spherical mirror, which then bounces light onto the silvered spot on the lens, back down the tube and out through a hole in the primary mirror to a focus outside the rear of the mirror cell. Maksutov-Cassegrains use much the same light path, the principal difference being in the curvature of the front lens or meniscus of the system. In such a manner, the light path is quite long in a relatively small instrument, as the light path makes three trips around the system, resulting in a long path and a larger focal ratio, generally between f10 and f13 or more.

The advantage of such a telescope is its portability, small size and stability coupled to its ability to provide a ready platform for an SLR or a CCD device. The small field of view ensures that any extended object fills the field, which is typically about 0.5° and makes looking for small objects easier. Due to the larger relative size of the longer focal lengths of catadioptric telescopes, Wratten and other filters for DSLR and CCD imaging are very useful and cut down on the glare and light transmission through such 'scopes without losing the resolution or fine detail in the object under observation.

Catadioptrics come in a range of apertures, from small 90-mm 'scopes to giant 40-cm telescopes, which are not really portable objects at all. Several well-known manufacturers make catadioptric telescopes, and they are available at reasonable prices in outlets worldwide. They are easy to set up and maintain, though use of a dew shield is a must, as the primary lens of the system is right at the front of the 'scope and will suffer from dewing if it is not properly shielded. Dew shields, dew zappers and associated equipment can be obtained from all good astronomical suppliers, or can be home-made quite easily and cheaply.

Eyepieces

There is no point in having a quality telescope and then spoiling the ability to see detail by using inferior quality eyepieces. It is often the case that a good telescope and mount can be purchased as a complete item along with eyepieces, only to find that the eyepieces are not the kind of quality that matches the telescope. Over the years, many amateurs have learned this lesson; the supplied eyepieces with any system are not as good as they could otherwise be, and they have invested in a range of eyepieces for use with the variety of 'scopes they may possess. Quality eyepieces are a must no matter what the telescope in use.

In visual observing the Wratten filters are screwed to the eyepiece or gently placed over the front element.

Like a telescope, an eyepiece will come in a choice of focal lengths. These are usually displayed on the barrel as 32, 25, 20 mm and so on. These figures give the user the focal length of the optical elements within the eyepiece. Even if the observer has built up a good standard set of eyepieces, each will perform slightly differently on each different telescope that the observer uses. The main denominator will be the difference in field of view and magnification of the eyepiece on each telescope. Magnification can be obtained with the following equation:

$$M = (f_t / f_e)$$

where f_t is the focal length of the telescope and f_e is the focal length of the eyepiece. The magnification is simply obtained by dividing the focal length of the telescope by the focal length of the eyepiece For example, a refractor may have a focal length of 1,000 mm. If the eyepiece of choice is a 20-mm then the magnification of such in use with this refractor will be (1,000/20) = 50 times, or a magnification of 50×.

If the 20-mm eyepiece is used with a system having a longer or shorter focal length, then adjustments have to be made accordingly. For example, a 700-mm focal length will now have a magnification of 35× (700/20), while a 1,400-mm focal length will now have a magnification of 70× (1,400/20). This magnification will also impact upon the field of view of each instrument using that eyepiece.

When choosing different focal length eyepieces it is important to remember two things—the size of the exit pupil and the necessary eye relief of each one. The exit pupil is defined by dividing the diameter of the primary lens or mirror of your telescope by the magnification of the eyepiece and is always quite a small number. The exit pupil of a 100-mm f10 refractor telescope using a 20-mm eyepiece is therefore 2 mm (100/50) and provides an adequate exit pupil for most observers. The secondary consideration when choosing an eyepiece is the eye relief given to the observer, especially for those who may wear eyeglasses, but as a general rule the smaller the focal length of the eyepiece, the smaller the exit pupil and therefore the more difficult in practice the eyepiece becomes to use. This can be alleviated choosing eyepieces that have an ultra-wide flat field, making it a little easier to find the image. A good eye cup on the eyepiece may also assist those that wear glasses as some can be either extended or folded back to enable the user to find a comfortable distance.

Most observers use multi-element eyepieces that have weird and wonderful names such as Plossl, Erfle and Orthoscopic; these eyepieces are now the standard for observers and give exceptional views through most instrumentation, giving flat fields with no astigmatism or blooming of images at the edge of the field. They also vary in price, and the super high quality types may cost as much as small telescopes themselves! Nevertheless, a selection of good eyepieces giving a range of magnification is a must for all astronomers and provide a bonus for seeing those small, faint or difficult objects due to variations in FoV, contrast and color rendition.

Most standard eyepieces have fields of view (FoV) that vary from 30° to 50°. Of course, when in use this field is not 30–50° at all. One has to now divide the magnification the eyepiece gives on a particular instrument into the field of view quoted. So, for an eyepiece with a FoV of 50° and a magnification of 50×, the apparent field of view through the eyepiece will be 1° (50/50) or 60 arc minutes, whereas one with a magnification of 50× and a FoV of 30° will give a FoV through the telescope of 0.6° (30/50), or a FoV of 36 arc minutes.

An additional tool for determining the FoV is to put a bright star such as Altair (α Aquilae) or σ Orionis on the edge of your eyepiece field and allow the star to drift through the field until it reaches the other side. Time this event and multiply the time by 15 to obtain the FoV in minutes and seconds. As 60 min make 1°, it will be easy to determine FoV with this method and adjust accordingly.

Observers will have to make all these calculations across the range of their 'scopes and eyepieces so that they will then understand the FoV through each instrument and will be able to use this information in observing the field for faint objects and making a mental determination of how large an astronomical object would appear in such a combination. This is extremely useful for hunting down faint extended nebulae or compact objects such as some planetary nebulae, as the FoV will enable the observer to gauge the size of the target in the eyepiece more effectively.

One additional accessory that will aid with magnification is the Barlow lens. This negative focus lens effectively doubles or triples the magnification of each eyepiece, and the quality versions have a demountable lens that will fit into the adapter sleeve of a DSLR camera, thus increasing the focal ratio of the optical system and enabling a smaller FoV around the object one wishes to image. These lenses are available from all good suppliers, but the quality can vary.

A technique that is very useful when using colored filters at the eyepiece (or any other for visual observing) is that of "blinking" the field, which involves *not* screwing the filter to the eyepiece barrel but holding it between the eye and eyepiece and moving it in and out of the field. Working like a blink comparator, the use of the filter then turns the details on your chosen object "on and off" as it crosses the field of view and makes the body stand out in sharper relief as some details gain clarity and the brighter ones fade, and the contrast brings out details that would otherwise be lost in the glare. This technique is easy to use and can make a clear difference to the observer's hunt for faint details, and it also works for some narrowband filters such as the OIII, which shall be the subject of later discussion.

Quality eyepieces married to a good telescope of reasonable aperture can be used to make important observations of Solar System objects, the Moon and some deep sky objects, too. Marrying a CCD camera to such systems can broaden the range of filters used beyond the Wratten system, but that is something we shall cover in a later chapter.

Conclusion

It is hoped that readers are convinced not to ignore the strange colored filters that may accompany their telescopes. If you don't yet have colored filters then it is recommended getting at least a selection of them from the earlier list, as they do proffer a new experience when gazing at the Moon or planets. The enhancement of detail and the reduction of glare should be ample inducement to own some filters that can be purchased inexpensively from astronomical suppliers. Most offer a selection as part of a lunar and planetary set wherein Wratten numbers 8, 11, 21 and 82A are the most common combinations in a set of filters for this kind of observing activity. Retailers will sometimes offer two sets, a "light" set, as noted with the Wratten numbers above, and a "dark" set, with Wratten numbers 15, 25, 47 and 58 to offer additional contrast and variety in your Solar System observing.

The table below gives a précis of the filters described above and their characteristics. Married to a good quality telescope with an appreciable aperture and good eyepieces the astronomer is well equipped to explore faint and subtle details in many objects.

Wratten no.	Color	Characteristics
25A	Red	Longpass filter blocking light short of 580 nm
23A	Light red	Longpass filter blocking light short of 550 nm
21	Orange	Longpass filter blocking light short of 530 nm
8	Light yellow	Longpass filter blocking light short of 465 nm
12	Yellow	Longpass filter blocking light short of 500 nm
15	Deep yellow	Longpass filter blocking light short of 510 nm
11	Yellow–green	Color correction filter, not a longpass filter
56	Light green	Color separation filter
58	Green	Color separation filter
82A	Light blue	Warming filter. Raises the intensity of long wavelengths
80A	Blue	Color separation and warming filter
38A	Dark blue	Absorbs red, green and UV light
47	Violet	Color separation filter

Colored filters are therefore highly recommended for Solar System objects and coupled with large aperture telescopes can reveal fine detail in the cloud belts and rings of the gas giants. It cannot be stressed enough that Wratten filters are an essential component of the lunar and planetary observer's toolkit. Let us then explore their use further by looking at the Moon and planets in detail.

Further Reading

Beisch J. Observing the planets with coloured filters. http://alpo-astronomy.org/mars/articles/FILTERS1.HTM

Muirden J (1993) The sky-watchers handbook. W H Freeman publishing

Thompson J. Bandpass filters for visual and video astronomy

Chapter 3

Observing the Moon with Filters

Anyone who has looked through a telescope is overwhelmed when they get their first sight of the Moon. It is difficult to compare the rich detail of its surface with any other object in the night sky. When one delves into the historical background of lunar observation, it is incredible and yet unsurprising to find that the first telescopic observers were so unprepared for what they would find on the surface that their maps bore little resemblance to real features. Their perceptions and expectations were influenced by those of the ancient Greek philosopher Aristotle, who took the view that the heavens were perfect—immutable and unchanging—and, thus heavenly bodies would appear to be lustrous and smooth. This view was passed down over the centuries, becoming incorporated into theological and social dogma.

So unearthly and anti-Aristotelian were the views through the telescope that the impact was felt throughout the literary and scientific world. Galileo announced the new landscape and a new philosophy via his tract *The Starry Messenger* (Siderius Nuncius). So shocked by the revelations of the telescope and the unpreparedness of people for this new world that when asked to comment on the view of the lunar surface the Welsh landowners Sir William Lower and John Prydderch at Treventy, using a crude telescope in 1609, merely commented that it "resembled a tart my cook made me last Sunday; here a bright vein and here a dark thing and all confusedly all over." Over the next century maps of the Moon grew from Thomas Harriot's first crude drawings to the beauty and cartographic excellence of Johannes Hevelius' *Selenographia*. The telescope introduced whole new worlds to the human eye and none so detailed as our nearest neighbor in space.

Since Galileo's time, the lunar landscape has been interpreted and re-interpreted cartographically until, with the advent of spacecraft, we have superbly accurate and detailed maps. However, for a novice, the complexity of the lunar surface can be

M. Griffiths, *Choosing and Using Astronomical Filters*, The Patrick Moore Practical Astronomy Series, DOI 10.1007/978-1-4939-1044-1_3,

© Springer Science+Business Media New York 2015

bewildering. A 150-mm telescope can reveal craters down to about 1.5 km across and linear formations that are even smaller; in fact, almost a million features are visible with such an instrument. Also confusing is the changing appearance of the features caused by the lunar phase, the phenomenon of libration and the slower nodding of the Moon's face known as nutation.

What, then, is the best way to start the seemingly monumental task of learning the features on the Moon, and how do filters assist the observer? Studying the progress of the lunar day and how the phase changes the appearance of all lunar features is the starting point. Observing such will familiarize even the modestly equipped observer with the changing lunar landscape. Introducing filters into your observations ensures that you gain the maximum input while reducing the glare of the sunlit Moon. Some filters enhance features in addition to reducing light transmission, so a good set of lunar and planetary filters is a useful adjunct to your observational armory.

After your initial amazed observation of the brilliance of the Moon and the incredible number of craters and other features on its face, you may next notice that the general visibility of these features varies according to two factors. One is how far the feature is from the line separating day and night on the Moon. The other is how far the feature is from the edge of the disk of the Moon.

The line separating day and night on the Moon is known as the *terminator*. At any phase except full Moon and new Moon, you can view the terminator with a telescope and see in a moment that the topographic features near the terminator all appear much sharper and more prominent than those far from it. The reason for this is the angle of sunlight on the lunar surface. Parts of the Moon far from the terminator are illuminated by the Sun from high in their sky. These appear washed out, revealing little relief due to there being no shadow—just as features here have shorter shadows when the Sun is overhead. On the other hand, parts of the Moon near the terminator experience a Sun that is low in their sky and thus shadows are long, picking out features as they undulate over the landscape. Even a very small or very shallow crater not visible most of the time is easily visible at such times because it stands out in strong relief.

The exceptions to this rule are lunar features that are not topographic—in other words, that do not represent variations in altitude. An outstanding example is the long streaks of brighter lunar dust and ejecta that radiate out from certain of the younger craters on the Moon. These "rays" are only visible when the Sun shines down from on high upon them—when they are far from the terminator. Near full Moon, several of the great systems of rays are impossible to overlook.

Almost as obvious as the effect of the terminator on a lunar feature's appearance is the effect of the feature's distance from the edge, or *limb,* of the Moon. You will see no round craters near the limb, because they look elliptical due to foreshortening from the angle at which we are viewing them. Some of the Moon's largest craters appear as little more than lines because we are viewing them at such a shallow angle.

Although lunar features are not always illuminated in the same manner, at least we should not have trouble recognizing them by their shape—more or less foreshortened due to their distance from the limb—if it were not for the fact that the amount of foreshortening changes. And of course that can only occur if the Moon is not keeping quite exactly the same face toward us but rather tilting its face slightly up and down, right and left. This libration enables the observer to see 59 % of the lunar surface, but also has the effect of making some features close to the limb easier to see at certain lunations and then difficult at others. Observe carefully the effect of libration on surface features near the limb. You can do this by studying a prominent feature near the Moon's limb for a number of nights and seeing how its apparent distance from the limb and its apparent shape both change. The largest feature good for this purpose is a small, isolated "sea," such as Mare Crisium. Even the naked eye can note its changes in apparent shape and position on the waxing Moon.

Lunar Landscapes

The Moon is the most detailed object available for users of modest equipment and one of the most incredible places affording exploration in the night sky. The *maria* or "seas" are obvious even to the unaided eye. Other lunar features are less conspicuous, but they are by no means less fascinating. The following information may be useful to enable you to observe and enjoy the many craters, maria, mountain ranges and mountains, rills and valleys, and faults and domes of the Moon. If you wish to examine the Moon in detail and print off large scale maps, then download the *Virtual Moon Atlas*, a free software program available at www.astrosurf.com that also provides a large store of information on geology, history and individuals after whom the features are named.

Craters

The Moon is absolutely plastered with craters, and they are without doubt the most conspicuous feature next to the maria. At full Moon some craters such as Copernicus, Kepler and Tycho have extensive ray systems, bright and extensive straight lines of light that cross the Moon as glowing trails of dust. The lunar southern highlands are a great place to examine craters; as the terminator crosses the surface during the phases, they can be discerned in ever changing detail.

The observer Ralph Baldwin has classified craters according to age—Class 1 are the new formations, sharp and bright with no overlapping craters and are aged between 2 and 3 billion years old. Class 2 are older than the mare eruptions that filled in the seas, and show signs of "softening" or erosion. Class 3 are older craters that show many signs of secondary impacts and erosion and are about 3.5–4 billion years old. Class 4 craters are almost as old as the Moon and were probably formed shortly after the Moon coalesced and that reveal extensive changes. Class 5 are also very old and have suffered lava floods that have filled their basins.

Examining craters with the attributes of age, secondary cratering or flooding by lava will help you get an estimate of the age of the lunar surface in that area.

This information can be supplemented by considering the geological ages of the Moon and how crater classes relate to these ages. The oldest features on the Moon are classed as *Pre-Imbrium* (4.5–3.9 billion years old) followed by the *Imbrium* period (3.9–3.2 billion years old). This period is broken down into three distinct ages; the *Nectarian* (3.9–3.8 billion years old) the *Upper Imbrium* (3.8–3.5 billion years old) and the *Lower Imbrium* (3.5–3.2 billion years old). This age is followed by the *Eratosthenian* (3.2–1.1 billion years old) and finally the *Copernican* (1.1 billion years old to present day). Using the *Virtual Moon Atlas* will enable you to identify craters and features that match these ages, and the materials you print off will aid identification in the more confusing areas.

Maria

The feature that is most apparent to the naked eye is the "man in the Moon," or the lava-filled basins and "seas" of the lunar surface. The idea that there was land and sea on the lunar surface stretches back to the Pythagoreans, who claimed that as a flawless sphere, the Moon simply reflected the light from the land and sea of Earth.

These areas are fairly ancient, between 3.5 and 2 billion years in age, and are the remnants of lunar volcanism. Unlike Earth, the Moon contains few volcanic features other than domes; the lava on the Moon extruded through cracks in the surface until eventual cooling of the lunar core occurred 2 billion years ago. Observing the maria under low angle illumination, one can discern some lava fronts, wrinkle ridges and domes on what otherwise appear with small instruments to be relatively smooth featureless surfaces. The maria divide into "seas, lakes and marshes" (Mare, Lacus, Palus).

One odd feature is the names of the seas. As the Moon waxes to 7 days, the seas have serene names such as Serenity or Tranquillity, while after 7 days the names change to intemperate ones—Sea of Rains, Ocean of Storms. This is due to the fact that ancient Moon watchers thought that the waxing Moon influenced the weather and that after 7 days of good weather would come 7 unsettled days! The Greek poet Aratus, a contemporary and friend of the philosopher Zeno, encapsulated this apparent weather influence in his poem "Diosemia" (forecasts) and subsequent lunar cartographers such as Hevelius and Riccioli kept this tradition.

Mountain Ranges

The mountain ranges of the Moon are mostly the borders of maria. Mare Imbrium, for instance, has for ramparts the Apennine, Caucasus, Alps, Jura and Carpathian mountains. All were thrown up like the walls of an enormous crater when the impact that caused the Imbrium basin occurred. Observers of the seventeenth and eighteenth centuries named the features after mountain areas familiar to Europeans. Much of their original nomenclature has been eradicated by time and bureaucracy, but some still remain intact. Although the Moon's mountains are much different than Earth's, they do compare favorably in size with all but Earth's tallest ranges, and are far bigger in proportion to the little world on which they are found.

The Apennines are the greatest, extending for 1,000 km with peaks up to 6,000 m high. They form a deformation of the terminator easily noted with the naked eye just after first quarter. The lunar Alps are best known for being cut by the marvelous Alpine Valley, a giant slash across the range that can be seen to good effect on the 6-day Moon. The Carpathians are not far from the magnificent crater Copernicus. This entire mountain chain is actually part of the buried rampart wall of Mare Imbrium. Other important ranges around the Moon include the Caucasus, Haemus, and Riphaean. Isolated mountains on the surface of the maria may form spectacular islands of light and, at the correct phase, cast dramatic long shadows. Pico and Piton and Tenerife are the best examples in the northern Mare Imbrium. Close to the great crater Plato can be found the Straight Range—60 km long, less than 2,000 m high, but very regular and discrete.

Valleys and Rilles

The greatest valley is the Alpine. It splits the Alps for over 110 km and is a dramatic dark mark in their bright midst around lunar sunrise on the 6-day-old Moon. Some valleys are really chains of small craters—for instance, the Rheita valley (this formation is 150 km long and 25 km wide, roughly). The term *rille* would generally be applied to a cleft narrower than a valley and is generally more sinuous in outline. Most rilles appear to be the result of lava tubes that have crossed the surface and then, over the aeons, experienced collapse. The most marvelous rille is Schroeter's Valley, running out from near the crater Herodotus and twisting dramatically, at one point broadening into what has been called "the cobra head." Also superb in a small telescope are the branching rille systems of Hyginus and Ariadaeus, which actually connect to each other, and that of Triesnecker. Reference to the *Virtual Moon Atlas* will enable the observer to locate these.

Faults and Domes

By far the most notable fault feature on the Moon is the Straight Wall—110 km long and over 300 m high with a fairly steep scarp slope. Much smaller are the mysterious lunar domes. These small surface swellings can be found in many places on the Moon. Two good areas for finding the domes are near the craters Arago and Prinz. The pre-eminent dome feature on the lunar surface is the large mountain Rumker in western Oceanus Procellarum. All these features can be found with the *Virtual Moon Atlas* and studied in relatively small telescopes.

Indeed, a small telescope will reveal a wealth of detail that can be captured relatively easily with modern photographic techniques.

To enhance the view through the telescope and to make the most of some photographic techniques, the use of filters will be a valuable addition to your lunar observing highlights.

Filters for Lunar Observation

Beyond the ubiquitous "Moon" filters sold with some telescopes, observers who wish to observe the exquisite detail of the lunar surface through a telescope will need to determine which filters are best suited to the purpose. Most of the really cheap Moon filters are simply neutral density filters that allow all wavelengths through but cut down on the light transmission. However, to bring out the best of the lunar surface, much more than merely reducing the light input into your telescope is necessary.

As the Moon is such a bright object, any telescope will resolve features that are not only small but have differences in shade and contrast. Many observers report that the best filters for observing the Moon are the yellow Wratten No. 8 and yellow Wratten No. 12. These two can also be augmented by the light blue Wratten 82A and the dark blue Wratten 38A. Many astronomers also recommend neutral density filters as an adjunct to these filters, and some have great success with a polarizing filter, too.

The effect of using these filters can be seen in Fig. 3.1a–c where we examine the Moon taken as an unfiltered and filtered photograph set:

One can notice from Fig. 3.1a that without filters the Moon is extremely bright and has plenty of bright glare that reduces contrast and thus visibility. Using a neutral density filter will cut down on the glare, as can be seen from Fig. 3.1b, but will only result in minimal contrast to the object, whereas the use of a blue Wratten 80A filter in Fig. 3.1c increases the contrast on bright objects and makes the mare stand out as darker, providing increased detail in the brighter ray craters and highland areas. The use of the correct filter can make a world of difference (literally!) to your visual acuity and in resolving the details on such a bright object.

In addition to the colored filters, a polarizing filter can make a great deal of difference to the contrast on the lunar surface. Some observers state that using a polarizing filter in addition to the usual colored ones can also result in increased contrast and thus detail, especially under high magnification with a large aperture telescope of 200 mm+. The results from using a polarizing filter can be seen from Fig. 3.2 here, where such a filter cuts down the glare appreciably and reveals a wealth of detail.

The use of yellow No. 8 and No. 12 filters also increase contrast and act almost like a pair of sunglasses poised over the eyepiece. The yellow transmission cuts down on the harsher blue areas and renders more contrast in the mare areas and in some craters, too. The rays are more attenuated through these filters, and better

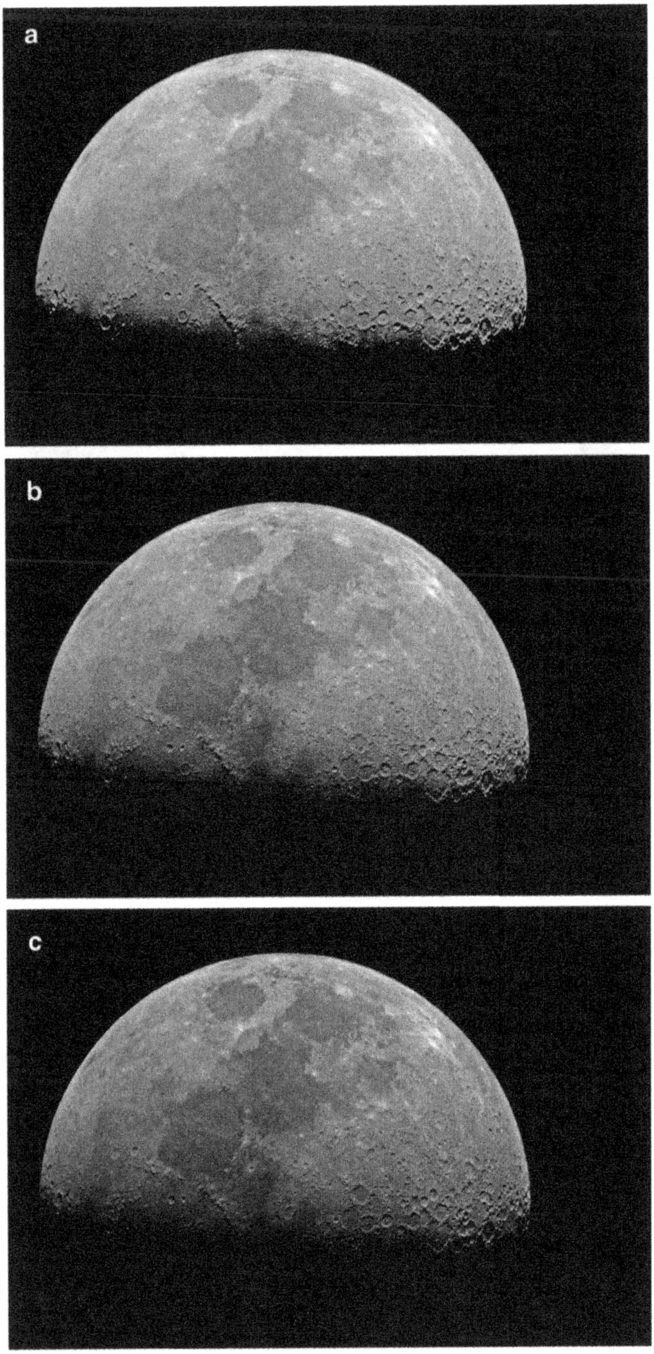

Fig. 3.1 (**a**) Bright glare moon. (**b**) Moon through ND filter. (**c**) Moon through blue filter

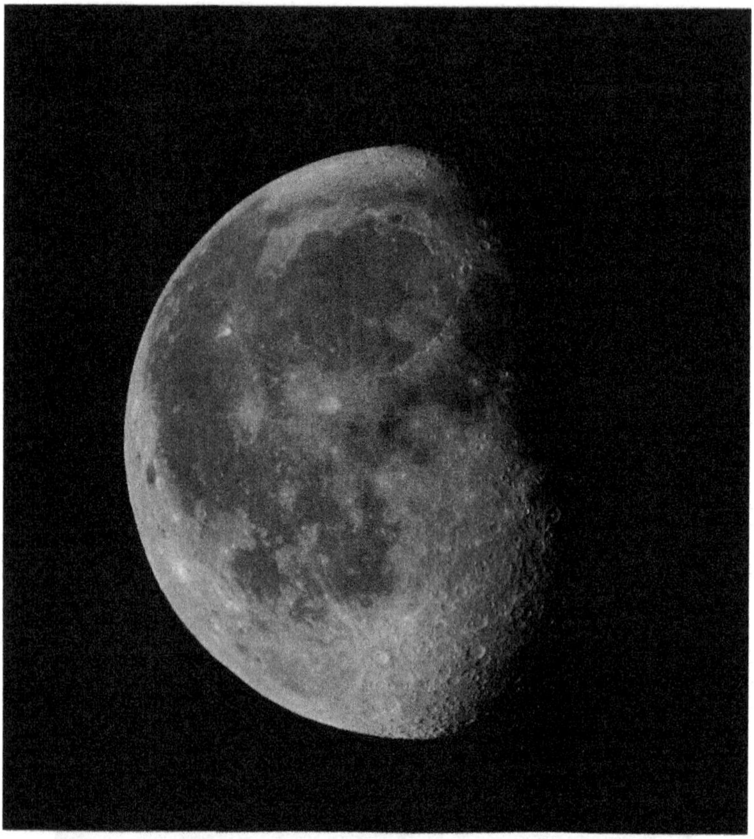

Fig. 3.2 The Moon through a polarizing filter

surface detail can be seen across the mare and highland regions with these longpass filters. This is not to say that neutral density filters or even the old "Moon" filters are of little use; indeed they can reduce the glare from the lunar surface and result in appreciable detail being seen through telescopes of relatively small aperture. As an illustration of this, note here that Fig. 3.3 was obtained through a 70-mm telescope with a Canon DSLR fitted with an ND 96 filter. Exposure was 1/500th second at f10 and an ISO of 800.

Obviously, a wide variety of filters can be used for lunar observing. Let us now turn to the Moon and examine a few features that can be seen in greater detail if filters are used. If the reader would like more information on lunar observing, this author would recommend any of the books written by Grego (2005).

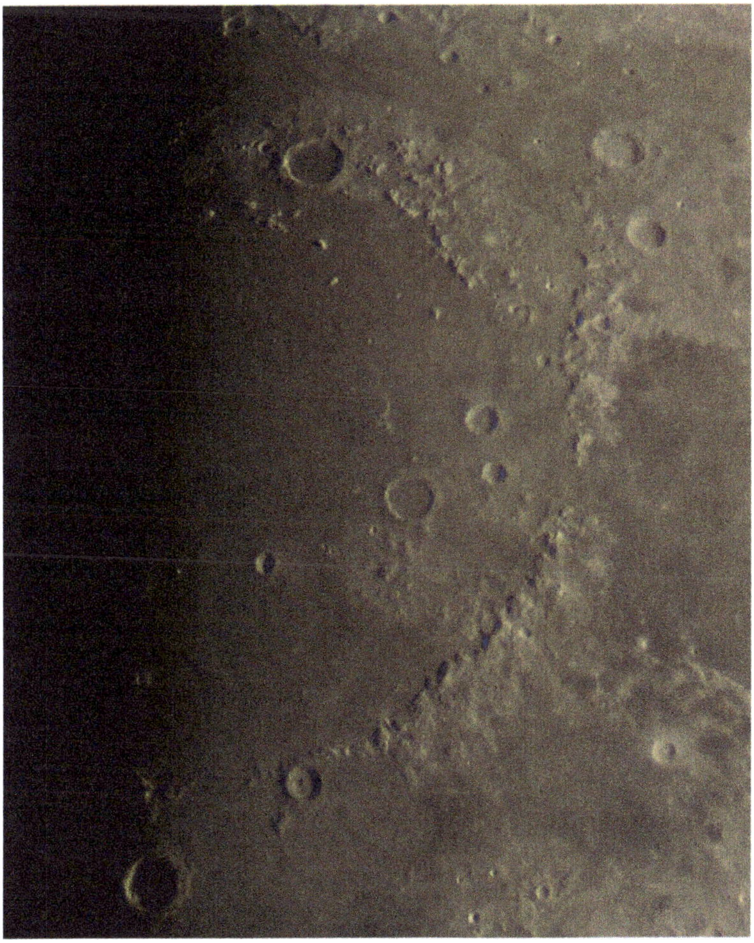

Fig. 3.3 Mare Ibrium, Alpine Valley and Appenines. ND filter

Phases of the Moon

New Moon to First Quarter

If you understand the basic astronomical factors such as phase, libration and nutation affecting the visibility of features on the Moon, you are ready to start learning individual features. Perhaps the most obvious plan for taking a first tour is to harness the lunar phases themselves by following the terminator across the Moon's face during the course of a month.

It is easier to note a selection of some of the most conspicuous lunar features as you makes progress in your observations. The schedule by which we observe them is day by day in the course of the Moon's progression from one new Moon phase to the next—a *lunation*. The amount of time elapsed since the most recent new Moon is called the Moons "age." Remember that the exact age at which the terminator reaches lunar features varies somewhat from month to month because of libration. This is yet another reason why you should record your observations, possibly including photographs of the entire crescent at low magnification and of especially interesting regions at high magnification. Utilizing a DSLR or DMK camera along with a small telescope may give you a good photographic record of the scene.

Two Days Old

Although the Moon can be observed far "younger" than this age under favorable conditions, even a 2-day-old Moon is often so low that "seeing" is likely to be poor. Most experts suggest studying this area near the eastern limb of the Moon just after full Moon (when the sunset rather than the sunrise terminator is moving across it) because the Moon can then be observed high in the sky for much of the night. The outstanding feature at this time is Mare Crisium (Sea of Crises). Mare Crisium is the only major mare on the near side of the Moon that is completely separate from the rest of the maria system, and it is completely surrounded by bright highlands. Well south of Crisium around this time, the terminator is passing the north–south series of the giant craters Langrenus, Vendelinus and Petavius. A blue Wratten 80A may assist in reducing some of the atmospheric dust and redness associated with sunset, though the No. 8 or No. 12 yellow filters may add a little more contrast.

Four Days Old

Both Mare Crisium and Mare Fecunditatis are now fully in view, and the Langrenus–Vendelinus–Petavius sequence is well seen. Small but bright Proclus lies just west of Crisium and has crossing rays, which under a high Sun are among the most prominent on the Moon. It can be seen as a bright spot on the accompanying photo. The craters Atlas and Hercules form a superb pair; Atlas is somewhat larger. Note Fracastorius, which was once a crater and now is a "bay" on the edge of Mare Nectaris. Near the southern horn of the crescent are the crater Rheita and the so-called Rheita Valley—which is not a true valley merely a chain of small craters.

Look out for the spectacular Altai Scarp, which appears as a long sinuous feature in the southern highlands. The scarp is probably a remnant of an Imbrium-type impact very early in the Moon's history. The 4-day Moon as seen and photographed

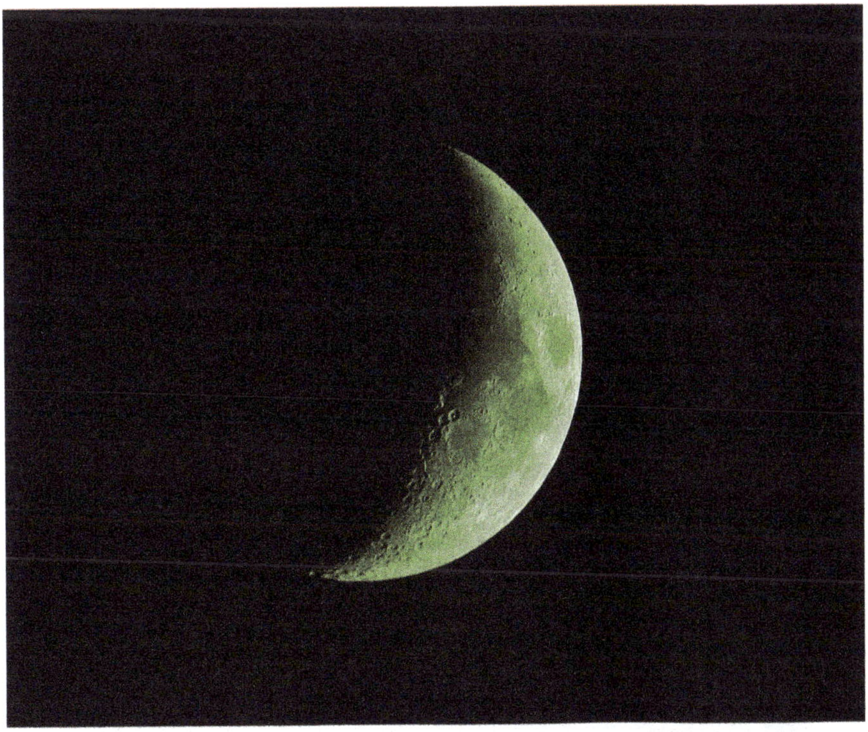

Fig. 3.4 A 4-day Moon

through the Wratten 12 yellow filter can be seen here as Fig. 3.4. Note the craters Theophilus, Cyrillus and Catherina around the edge of Mare Nectaris and the increased contrast this filter provides on some features. Theophilus, one of the finest craters on the Moon, has walls reaching to 5,500 m above its floor and a prominent central mountain that is well seen during the fourth and fifth lunar day. It overlaps the older, less well-formed Cyrillus and with Cyrillus and Catharina forms a spectacular western border to Mare Nectaris. The Altai Scarp stretches away to the southeast and should be quite prominent as a bright wavy "line." The light blue Wratten No. 82A will also assist at this stage in seeing some of the details in these craters and in picking out the edges of the Altai Scarp.

Six Days Old

Now all of Mare Nectaris and Mare Tranquillitatis is lit, as is most of Mare Serenitatis. The craters Aristoteles and Eudoxus are prominent neighbors in the highlands between Serenitatis and the east end of Mare Frigoris. Posidonius is prominent on a kind of cape at the eastern edge of Serenitatis; Plinius is visible in the strait between Serenity and Tranquillity. Near Plinius is bright Menelaus; similarly small but bright Maskelyne, named after the fifth Astronomer Royal and the villain of the longitude story, is close to the entrance from Mare Serenitatis to Mare Nectaris. South of Sinus Medii is a chain of prominent craters, among the best on the Moon. These are, in decreasing order of size, Ptolemaeus, Alphonsus and Arzachel, which form a chain. Compare their appearance. Ptolemaeus has a smooth floor filled with lava flows, while Alphonsus and Arzachel have the typical rampart walls and central peaks. The light blue Wratten filter No. 80A is a must to use here to reduce the glare and increase the contrast in the southern highlands.

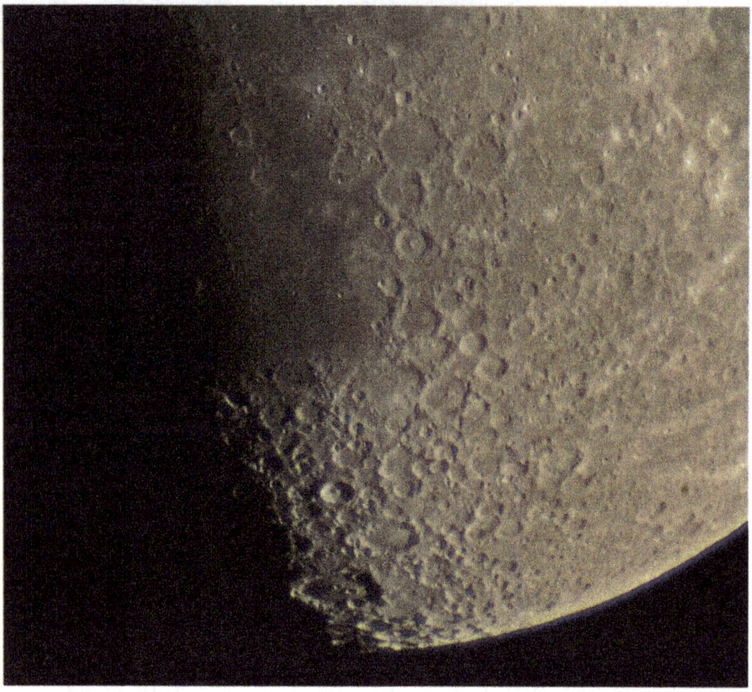

Southern Highlands - Neutral Density filter

Seven Days Old

Commonly known as *first quarter*, we can view a host of superb craters and mountains near or on the terminator. The Moon's most spectacular mountain range, the Apennines, are partly in view, and the wonderful crater Eratosthenes, named after the Alexandrian librarian who first made an accurate measurement of Earth's circumference is well seen at their southern border. Within the next day or so, the southwest end of this range will catch the sunrise, forming a peninsula of light that even the naked eye can see as a deformation of the terminator. A break between the northeast end of the Apennines and the beginning of the Caucasus Mountains is a strait between Mare Serenitatis and the still mostly dark Mare Imbrium.

Just to the Imbrium side of the strait, the craters Aristillus and Autolycus form a fine pair. A little farther north, in the Alps of the Moon, the impressive Alpine Valley is now seen as a less prominent cut of shadow through the mountain range, as it was during day four. The southern highlands are thick with major craters, and now is the time to see two huge but worn neighbor craters that are difficult to identify near full Moon—Hipparchus and Albategnius. Glance around the eastern border of Mare Imbrium and examine the lone mountains Pico, Piton and Tenerife. Note the long shadows these mountains cast. The blue Wratten 80A or the 82A can be used to great effect here to increase contrast and increase detail on the lunar surface.

Eight and Nine Days Old

The few days after first quarter are a busy time for lunar observers. In no other period do so many spectacular features come into view. The terminator is crossing the large extent of Mare Imbrium (Sea of Showers). The southeastern rampart of Imbrium is the Apennines, whose whole curving line is now visible. The range points westward to what is generally considered the most magnificent crater on the Moon, Copernicus. Often termed the "monarch of the Moon," Copernicus is magnificent and rich in detail and repays using either a Wratten No. 56 green filter or the darker blue 38A Wratten to bring out these exquisite details.

As the terminator glides past the crater, you can admire the great height of its walls, the marvelous central triple-mountain mass and the remarkable detail in its rough floor. Northwards, in contrast to the brightness of Copernicus is the dark floored majesty of Plato, a large crater in the highlands separating Mare Frigoris from Mare Imbrium. Plato is interesting in that its floor looks so smooth and regular to users of small instruments. Just to the south and east of Plato are the individual mountains, Pico and Piton, and the line of peaks called the Straight Range; Archimedes, Aristillus, Autolycus, and Timocharis are prominent craters within Mare Imbrium, but are seen to better effect around the 8-day phase.

Also visible at this stage is Rupes Recta, the so called Straight Wall, a linear feature in Mare Nubium that repays use of a ND filter or a yellow Wratten No. 12. Figure 3.5 shows this lovely hair-like feature that reverses to become a bright line during sunset over this area at around 21 days of the lunar phase taken through a neutral density filter.

On the 9-day-old Moon, the terminator has already swept past Sinus Medii (Central Bay), which marks the center of the Moon's face. Farther south are still

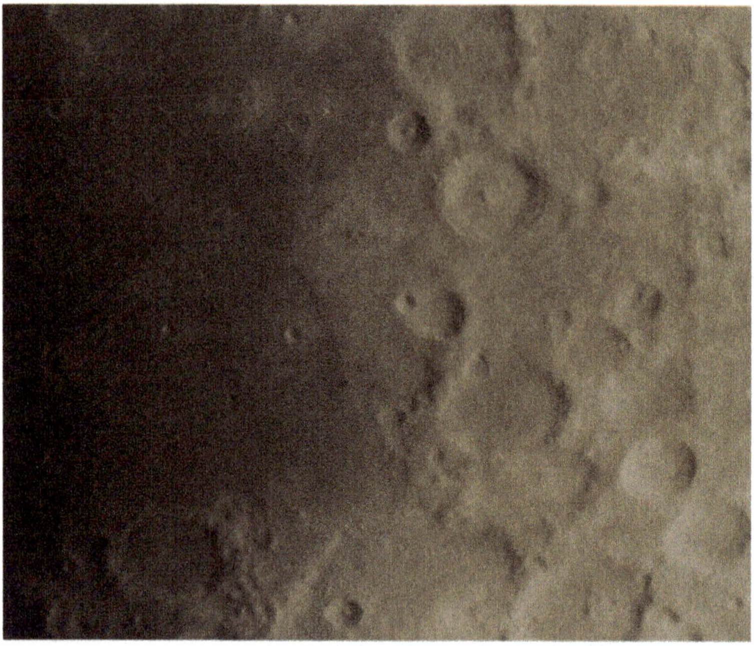

Fig. 3.5 Rupes Recta

more great craters near the terminator—among them the famous, young and beautifully formed Tycho and the foreshortened, old, but enormous Clavius. Most of Mare Nubium is now in view, with the crater Bullialdus very prominent at the sea's northwestern edge.

Ten Days Old

The fabulous Sinus Iridum (Bay of Rainbows) is at the terminator. Under low angle illumination, several wrinkle ridges can be seen, almost looking like waves as they enter the bay. On the bay's western border, the Jura Mountains catch the Sun before the lower land and therefore shine out like a "jeweled handle" in darkness. At the southern extremity of Mare Imbrium, the full length of the Carpathian Mountains, northwest of Copernicus, is now lit, forming a partial boundary between Mare Imbrium and the even vaster but irregular, "overflow" mare, Oceanus Procellarum (Ocean of Storms). For several days, the great ray systems of the eastern part of the Moon, those of Proclus and little Stevinus, have been prominent. From this time forward, the ray systems of Copernicus and Tycho begin to glowing ever more prominently until at full Moon they completely dominate the lunar face.

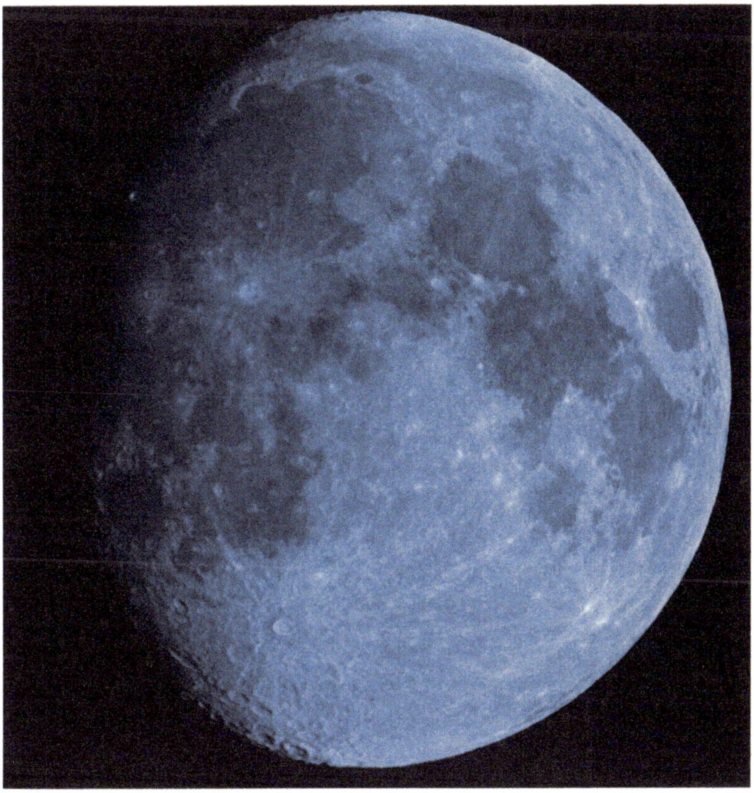

Fig. 3.6 A 10-day Moon through a Wratten 80A

This is the time to start darkening the lunar surface considerably by using some of the blue filters such as the No. 38A and even the No. 47 violet, depending on the aperture of your telescope. The lunar surface is very bright, and use of the blue filters will increase contrast considerably. This can be illustrated in Fig. 3.6 where a blue 80A filter was used to reduce glare through an 80-mm telescope. This is a DSLR shot using a Canon 30D at ISO 800 with an exposure of 1/600 second at f6.2.

Eleven to Thirteen Days Old

The wonderful, bright companions of Copernicus in Oceanus Procellarum, the craters Kepler and Aristarchus, are now visible. Aristarchus has the greatest surface brightness of any place on the Moon, and both it and Kepler possess superb ray systems. Very near Aristarchus is contrastingly dark Herodotus, from whose northern wall runs and winds for a long distance the lovely Schroter's Valley, perceptible even in quite small telescopes. By now the lunar surface is reflecting so much sunlight that it becomes difficult to separate features in the glare. South of Kepler and Oceanus Procellarum, the entirety of the small Mare Humorum is visible, with the huge crater Gassendi on its northern edge, although the sea is best seen at 11 days old.

Located down in the southwestern part of the Moon, far below Humorum, Schiller is an elongated feature that apparently resulted when the walls separating two craters broke down. Not far away from Schiller, Schickard is a vast, worn crater. Close by is the remarkable crater Wargentin that has become "overfilled" with lava, testimony to the volcanic nature of many lunar features. To the south of Schiller, near the very limb of the Moon, we should now try glimpsing the ancient, worn outline of the largest formation called a crater, or "walled plain," on the near side of the Moon—Bailly, about 300 km across. Even at the most favorable librations, Bailly is so near the southern limb as to be a difficult sight. Try using different filters here to increase the contrast and reduce glare so that these features reveal a little more detail under conditions of good seeing. Whichever ones the observer uses will be useful, but the darker green and blue No. 56 and No. 38A Wratten filters usually produce more contrast on the bright lunar surface.

The Full Moon

There is little contrast on the shadowless full Moon. This is a time to enjoy the glory of the great ray systems, especially that of Tycho. The ray that bisects Mare Serenitatis may be derived from Tycho, over 1,500 km away. Now is also the time to get the best look at features near the Moon's western limb. The darkest floor of any crater is that of Grimaldi, which is rivaled by that of nearby Riccioli. Finally, very near the northwestern limb, a formation perhaps even larger than Bailly can be glimpsed. It would appear that this crater, Otto Struve, seems to be two craters not well seen due to the angle and unfavorable librations.

During a favorable libration, try spotting the great mountains of the Mare Orientale in the southwestern quadrant of the full Moon. All that can be seen is a jagged edge to the limb, though on occasion it is possible to see into the plains of the great basin beyond these mountains. This area of the Moon was little known before the advent of spacecraft, and amateur observations of these libration zones were highly important to lunar research.

As the Moon wanes from full, rising later and later in our evenings, the sunset terminator on it slides over from lunar east to lunar west, repeating the high-contrast visibility of the first 14 days of features just described. But there are some subtle differences. One of the most beautiful is what happens to the Straight Wall. Appearing as a dark line after first quarter, the Straight Wall reappears around last quarter as a lovely scratch of brilliance, as its face, pointing to lunar west, catches the full sunlight. Note also on the waning Moon the fading of the previously bright ray systems. By now, observing the Moon is an activity that requires perseverance through the night or rising very early to observe the transitions across its surface and of course the use of filters to see the differences that the low-angle illumination of the Sun provides features that are now lit from a different direction.

During the full phase, increasing the contrast on the lunar surface is only possible by using deep colored filters or, even better, the polarizing filter so that the bright highland areas can be separated even more from the mare. Even a combination of ND and polarizing filters may cut down on the glare and make observing or photographing the lunar surface during the full phase an easier proposition and result in an image with greater contrast.

Lunar Eclipses

This phenomenon is well worth looking out for and is a celestial marvel that is both delightful and instructive. Unlike a total solar eclipse, wherein the Sun goes black due to its occultation by the Moon, a lunar eclipse can turn a lovely coppery red color as sunlight is refracted or bent through Earth's atmosphere. The color can also be intensified, or fade away to nothing, leaving an inky blackness where the Moon has been, depending on the amount of dust in the atmosphere. After the eruption of the Philippines volcano Mt. Pinatubo in 1991, the next lunar eclipse was very dark. Successive eclipses have returned to their usual color; the photograph here was taken by the author during the 2007 eclipse (Fig. 3.7).

Wratten color filters can be used during a lunar eclipse to augment the beautiful coppery red of the Moon during totality. The Wratten No. 21 orange and the

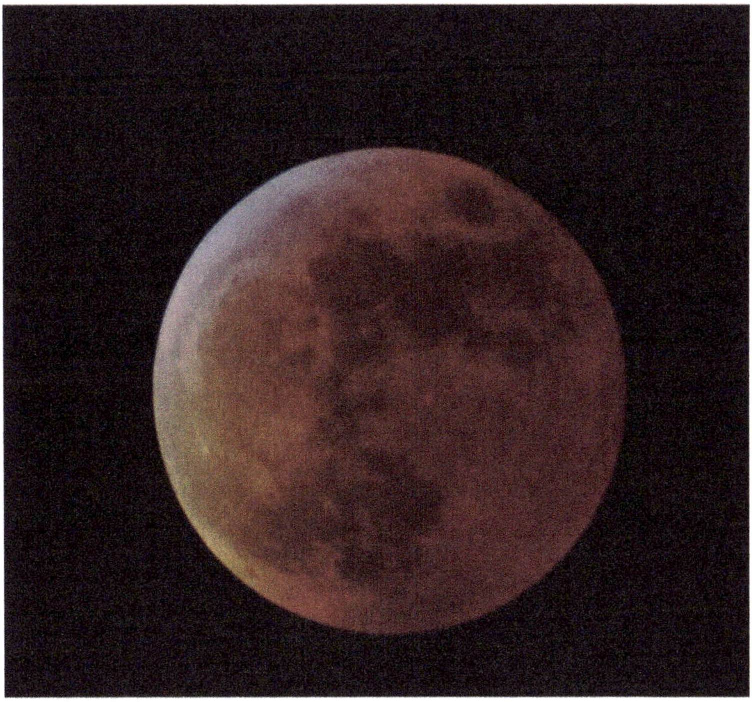

Fig. 3.7 Lunar eclipse

Wratten No. 11 yellow assist with this, though most observers prefer the color to remain neutral so that the Danjon scale of lunar brightness during eclipses can be gauged. For photography, though, the addition of a neutral density filter may enhance the photographs and reduce the contrast between the dark sky background and the Moon itself. In the above photograph an ND filter with a light transmission of 60 % was used. The red color of the Moon has certainly been added to by means of the filter.

The following table gives the observer a quick reference guide to features that are highlighted by means of filters.

Wratten No	Color	Features
ND	N/A	Lowers glare and increases contrast
8	Light yellow	Increases contrast and increases detail in ray craters
12	Yellow	Increases contrast on maria and highland areas
80A	Blue	Increases contrast on maria and across all bright ray and crater features
Polarizing	N/A	Increases contrast in rays and reduces glare

The following illustrations give the phases of the Moon and highlight some of the craters and other features that can be seen at those times on the waxing/waning Moon. Using the above filter sets should enhance your observations of these features.

Conclusion

The Moon is a wonderful object for personal exploration, and many observers spend a lifetime getting to know its surface in intimate detail. Modern tools such as the *Virtual Moon Atlas* and applications for iPad, iPhone or Android devices are excellent adjuncts that assist with your identification and observing pleasure. Photographing the Moon with such filters as described earlier will enable one to glean a complete record of the visible surface and identify features at leisure in addition to building up your confidence in using filters and providing valuable experience that can then be applied to other objects.

The reader has no doubt remembered from Chap. 2 that filters are used extensively to visually explore the bodies of the Solar System. It is to these planets that we now turn.

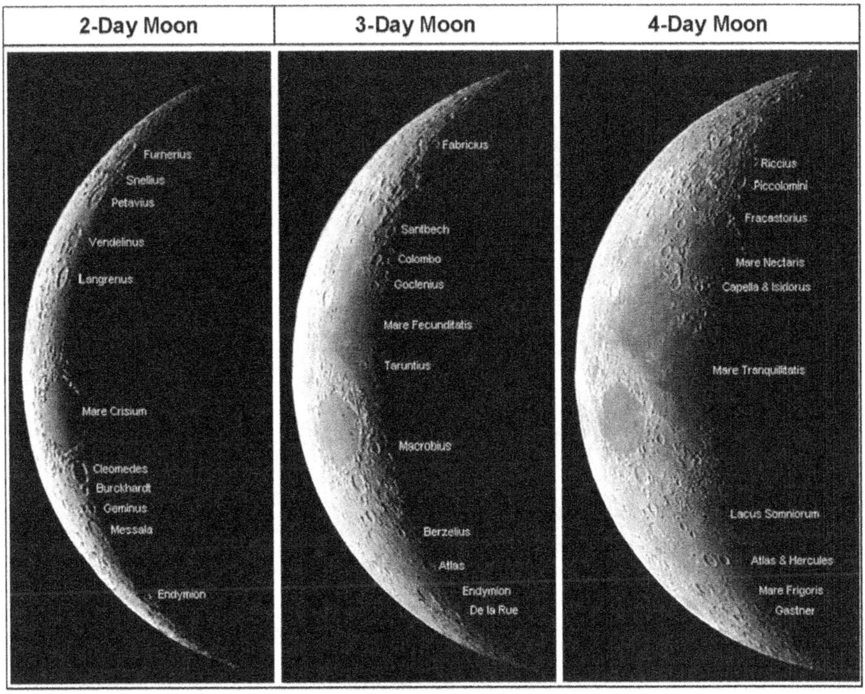
2-Day Moon
Furnerius
Snellius
Petavius
Vendelinus
Langrenus
Mare Crisium
Cleomedes
Burckhardt
Geminus
Messala
Endymion
3-Day Moon
Fabricius
Santbech
Colombo
Goclenius
Mare Fecunditatis
Taruntius
Macrobius
Berzelius
Atlas
Endymion
De la Rue
4-Day Moon
Riccius
Piccolomini
Fracastorius
Mare Nectaris
Capella & Isidorus
Mare Tranquillitatis
Lacus Somniorum
Atlas & Hercules
Mare Frigoris
Gastner

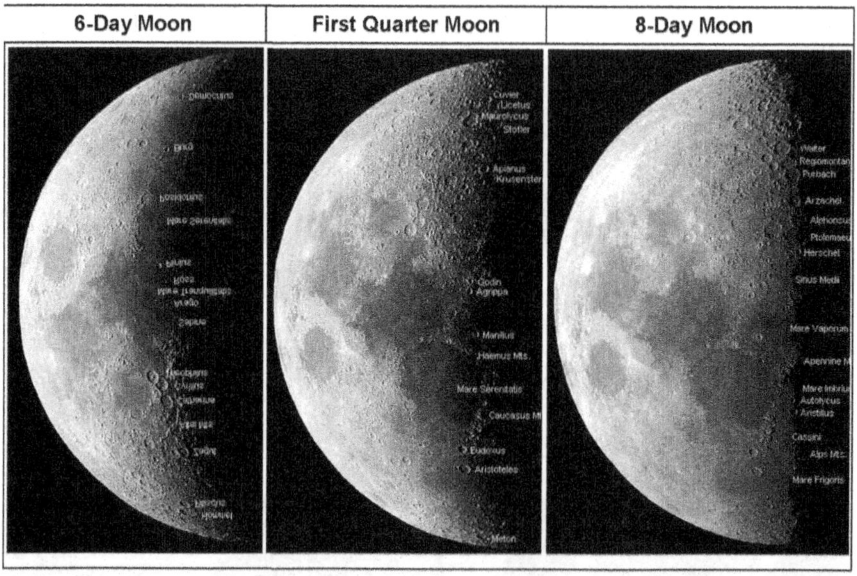
6-Day Moon
First Quarter Moon
8-Day Moon
Cuvier
Licetus
Maurolycus
Stofler
Apianus
Krusenstern
Godin
Agrippa
Manilius
Haemus Mts.
Mare Serenitatis
Caucasus Mt
Eudoxus
Aristoteles
Meton
Walter
Regiomontan
Purbach
Arzachel
Alphonsu
Ptolemaeu
Herschel
Sinus Medii
Mare Vaporum
Apennine M
Mare Imbriu
Autolycus
Aristillus
Cassini
Alps Mts.
Mare Frigoris

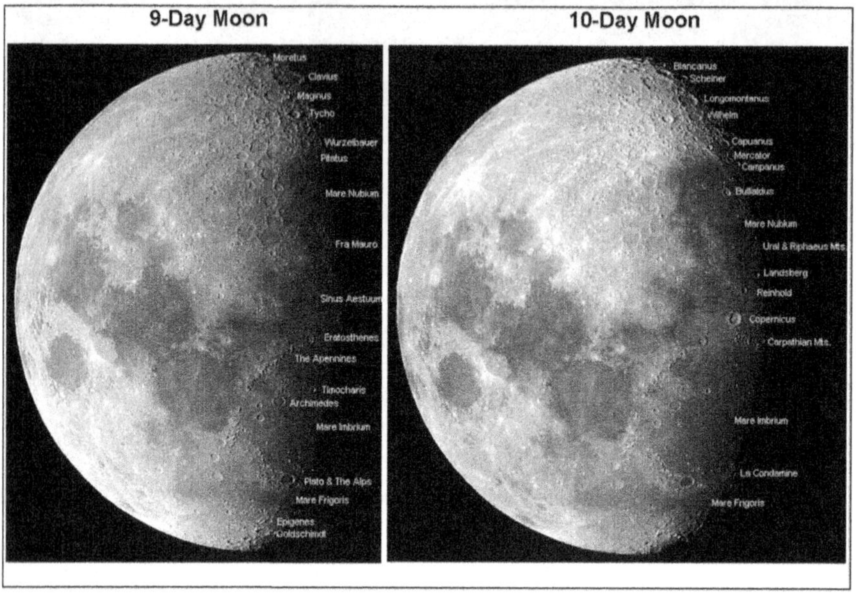
9-Day Moon
10-Day Moon
Moretus
Clavius
Magnus
Tycho
Wurzelbauer
Pitatus
Mare Nubium
Fra Mauro
Sinus Aestuum
Eratosthenes
The Apennines
Timocharis
Archimedes
Mare Imbrium
Plato & The Alps
Mare Frigoris
Epigenes
Goldschmidt
Blancanus
Scheiner
Longomontanus
Wilhelm
Capuanus
Mercator
Campanus
Bullialdus
Mare Nubium
Ural & Riphaeus Mts.
Landsberg
Reinhold
Copernicus
Carpathian Mts.
Mare Imbrium
La Condamine
Mare Frigoris

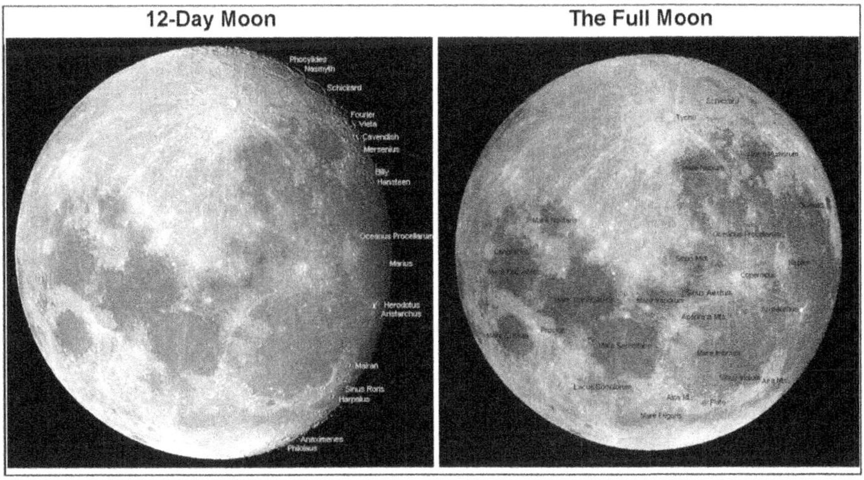

Further Reading

Cook C (2012) The Hatfield Lunar Atlas. Springer, New York
Grego P (2005) The Moon and how to observe it. Springer, New York
North G (2007) Observing the Moon. The modern observers guide. Cambridge University Press, Cambridge, UK
Wlasuk P (2012) Observing the Moon. Springer, New York

Chapter 4

Observing the Solar System with Filters

In Chap. 2 we briefly explored Wratten filters for visual observation. In this chapter we take each of the planets in turn and see how the observer can use these filters to enhance each body to the maximum.

Obviously the amount of detail that can be seen on a planet depends on atmospheric seeing and the size of the telescope in use. For observers using small telescopes in the 50–80 mm region, the best planets to observe will be either the closest or brightest. This covers Mercury, the gas giants Jupiter and Saturn and the bright planet Venus. Quite a lot of detail can be seen under steady seeing of both of the gas giants no matter what the aperture, though increased magnification and loss of contrast can result with a small telescope and high magnification. Just experiment until you can see the maximum detail with the best magnification and filters. With Venus, several filters may be used depending on the phase of the planet and the size and magnification of the telescope.

Notice that Mars is not included in the above list. Mars as a planet is not the easiest on which to glimpse detail, and the need for filters is at its greatest here to separate the subtle features of its surface. It is also a very small planet at a great distance and so demands on a larger aperture to ensure that the light transmission lost by using a filter is no hindrance to the observer. Apertures of 200 mm and above are recommended to get high magnification filters views of the Martian surface, though of course the observer can try and see general features with any aperture.

With regard to planetary observations, it is easier to either "blink" the filter across the eyepiece or place the eyepiece carefully on to the eyepiece cup so that it can be quickly removed and substituted for another filter without having the bother of screwing the filter to the eyepiece and then laboriously removing it every time.

M. Griffiths, *Choosing and Using Astronomical Filters*, The Patrick Moore Practical Astronomy Series, DOI 10.1007/978-1-4939-1044-1_4,
© Springer Science+Business Media New York 2015

Observers who wear eyeglasses may find that this enables a little longer eye relief, but it depends on the quality of the eyepiece.

The following sections will deal with general views of the planets through various filters and provide a table at the end of each section that sums up the filters and features for each planet.

Mercury: The Innermost Planet

Mercury is the most neglected of the planets. Sometimes brighter than Sirius, the brightest naked-eye star, Mercury is regarded by most amateur astronomers as being too close to the Sun to be an easy target, but it can be seen and appreciated nevertheless with some planning and preparation.

Mercury was once thought of as the smallest planet, the hottest planet—and also the coldest planet, due to the mistaken belief, based on the work of astronomers such as G. Schiaparelli and E.M. Antoniadi, that it always kept one face pointed away from the Sun. Several science fiction stories were set around this weird property, notably Hal Clement's excellent "Hot Planet." We now know that Triton, the large satellite of Neptune, at −235 °C, is the coldest body ever imaged in the Solar System, and Venus, with its extreme greenhouse effect, is the hottest. A distinction that Mercury will never lose, however, is that of the innermost and the fastest planet, orbiting the Sun in 88 days.

This is partly what makes it so elusive to observers. It is an "inferior planet" (closer to the Sun than Earth is), like Venus, and goes through the same order of major events: superior conjunction on the far side of the Sun, greatest eastern elongation in the evening sky, inferior conjunction on the near side of the Sun, greatest western elongation in the morning sky and then back to superior conjunction. But, whereas Venus can travel out to 47° from the Sun and set 4 h after (or rise 4 h before) the Sun at its best apparitions, Mercury can manage only 28°, usually setting a few minutes after the end of evening twilight (or rise a few minutes before the start of morning twilight). Mercury can never really be seen properly in a fully dark sky. This signals the planet as one that definitely requires the use of filters to increase its contrast against the twilight sky.

Mercury has a much lower surface brightness than Venus, and it always appears closer to the Sun and thus in a brighter area of sky. The observer will have to observe on very clear, haze-free days, when the Sun has no more than a small aureole of scattered light around it. The greater proximity to the Sun also means that you will have to be even more careful about not getting the blinding solar disk in your field of view, especially if you try scanning with setting circles. Actually, you are quite likely to find Mercury in daytime by using a GOTO mount accurately set.

Another observational problem is that the apparitions of Mercury are not all equally good for observers at a given latitude. Greatest elongations of Mercury vary between 17° and 28°. For an observer at mid-northern latitudes on Earth, the *minimum* greatest elongation of Mercury could be the year's best. What matters is the time of year because it determines the angle of the ecliptic as it intersects the horizon. At mid-northern latitudes, the angle of the ecliptic is steep around spring equinox in the dusk sky and around the autumn equinox in the dawn sky. So the best evening apparitions of Mercury thus tend to be those near the spring equinox; the best morning apparitions are those near the autumn equinox. At such times, a planet's elongation can be almost vertical at sunset; its angular altitude is greater and thus displays the planet to advantage.

A good apparition features Mercury as easily visible and setting over 1.5 h after the Sun (or rising this long before the Sun) around the time of greatest elongation. This means that Mercury can be seen about 2 weeks before and 1 week after a greatest evening elongation at a good apparition—so swift is this planet in its orbit around the Sun.

Filters for Observing Mercury

Actually, the surface brightness of Mercury is great enough that observers do not want to see it through the telescope against a fully dark sky. However, if you wait until after sunset to find Mercury in the twilight sky, the planet is so low that "seeing" is usually bad. One possible solution to this problem is to observe Mercury after sunset often enough to catch the few occasions when "seeing" is reasonably good even down low in the sky. Another is to observe Mercury high in broad daylight—which is achievable using a good telescope, one equipped with either good setting circles or a GOTO mount, but Mercury's features tend to get blanked out due to the excessive sunlight. To overcome this it is possible to use a Wratten No. 25 red filter, which then darkens the background sky and banishes the brightness of Mercury, enabling some surface features to be glimpsed under high powers.

Under such a filter, Mercury seems to take on more of the recorded pinkish tinge it generally displays. Figure 4.1 displays Mercury's general appearance as seen through a Wratten No. 80A blue filter. However, check the color of Mercury yourself and see what you think. If Mercury is higher in the sky and seen against a darker backdrop, then try observing it with a Wratten No. 12 yellow filter. This could separate some of the surface features by contrast and enable some darker "mare"-type areas to become visible in instances of good seeing. However, beware! A lot of this surface detail is in the eye of the beholder. Maps drawn by eminent astronomers such as E.M. Antoniadi at the beginning of the twentieth century turned out to have no recognizable connection to the maps made by *Mariner 10* in 1973 and certainly no connection to the surface as imaged by NASA's MESSENGER spacecraft. Most of these surface features could be just areas of different contrast connected by hand–eye and brain coordination to display something else.

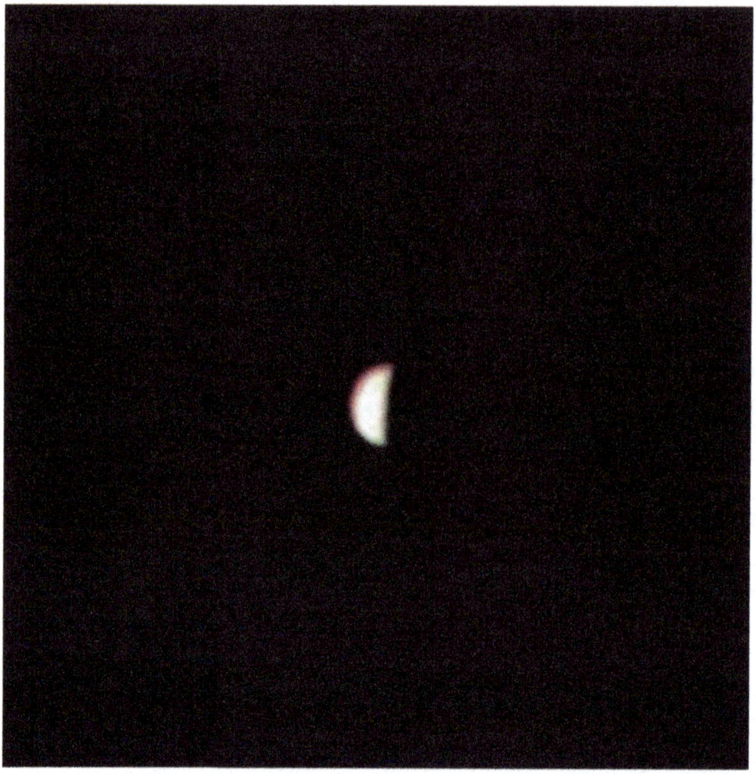

Fig. 4.1 Mercury with No. 80A filter

Mercury can be seen for more nights and more apparitions than the average amateur astronomer thinks. Trying to see Mercury at a fairly poor apparition, besides being a chance to test and sharpen your skills, might actually pay off in showing you features of a face of Mercury very rarely seen, especially if filters are used. Using a combination of Wratten No. 15 deep yellow or deep red Wratten No. 25 or even a light blue No. 82A on the bright twinkling star-like object at low elevation may make some features swim into view. It is imperative that the filters also be coupled to a high magnification eyepiece on a fairly large aperture telescope to gain the most from the clashing dark filter/twilight sky.

With these filters, the observer should immediately notice that Mercury does show phases. You will find observing Mercury when it is still almost fully lit far more difficult than doing so with Venus, for example, because Mercury then is still at such a small elongation. Observers who wish to see an almost "full" Mercury will be disappointed to find that under such illumination not only is the planet just above or below the Sun (depending on node), but that it is also on the other side of the Sun and is a much smaller object in angular diameter. If this type of observing is attempted then a Wratten No. 25 or even the deep violet Wratten No. 37 may show the planet, though any surface detail may well be lost.

Shortly after superior conjunction in the evening (or not long before superior conjunction in the morning), Mercury appears brightest and is therefore best set for observing with filters. However, Mercury's increase in apparent diameter as it approaches us on our respective orbits cannot make up for its dwindling phase for long. Mercury's phase shrinks rapidly in the week or two before inferior conjunction, and its magnitude dims into insignificance.

One time when filters of a different kind may be used is during transits of Mercury across the solar disc. White light filters or a Hα filter may be used to advantage at such times, though it must be pointed out that transits are relatively rare. These usually happen in November or May, and the following table gives the time in UTC of the next transits, but check on local times and visibility online.

Date	Start time	Mid transit	End time	Duration
May 9, 2016	11:12	14:57	18:42	7 h 30 min
November 11, 2019	12:35	15:20	18:04	5 h 29 min
November 13, 2032	06:41	08:54	11:07	4 h 26 min
November 7, 2039	07:17	08:46	10:15	2 h 58 min

Under such conditions the materials covered in the chapter dealing with solar filters will be useful. For those equipped with small telescopes, a simple Baader Astro filter paper will suffice to see the tiny dot cross the surface of the Sun, as can be seen from using such a filter during the May 2003 transit (Fig. 4.2).

So to sum up, the filters that are most useful to observe the planet Mercury with are shown in the following table.

Wratten No	Color	Features
No. 15	Deep yellow	Contrast features and separation of sky
No. 25	Deep red	Contrast features and darkening of sky
No. 80A	Blue	Contrast features and darkening of sky

Can the same filters be applied to the planet Venus, as it, too, is an inferior planet and displays many of the uncertain contrast features of Mercury?

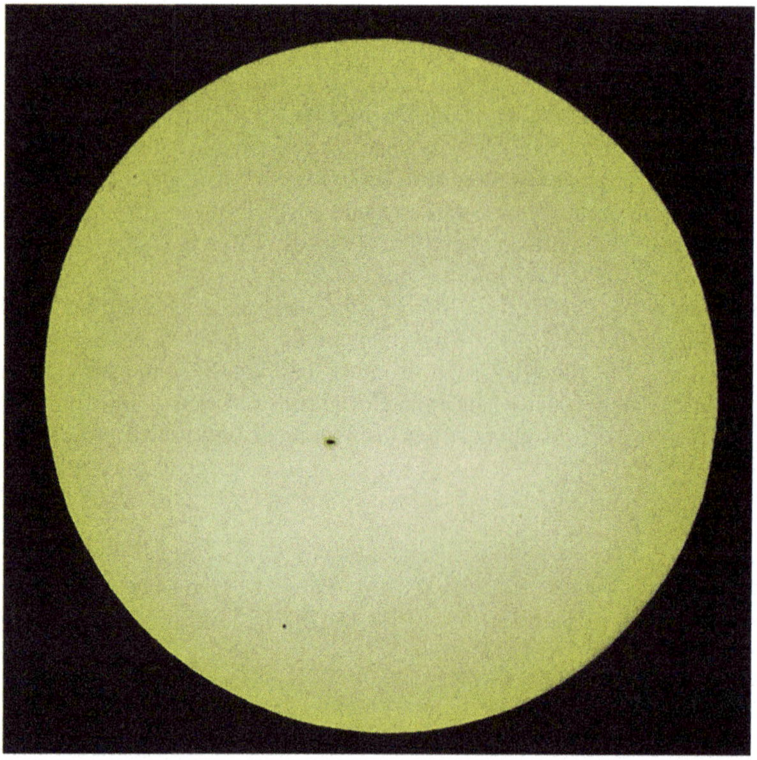

Fig. 4.2 Mercury in transit

Venus and Wratten Filters

Venus is the brightest planet visible to the naked eye and approaches Earth closer than any other planet. Through a telescope, Venus can also subtend a large visual angle and appear even larger than Jupiter; although at only 9 arc seconds at superior conjunction, it cannot be seen due to its proximity to the Sun. While orbiting in its closest approach to Earth, however, its slim crescent can grow to 68 arc seconds. Yet despite this, Venus is often a difficult planet to observe telescopically and usually shows less detail than Jupiter, Saturn or Mars. In fact, it is so difficult to discern detail on the surface that most astronomers, amateur or professional, do not bother to give it more than a cursory glance. This is partly due to its great brilliance; the planet usually shines between magnitude −3.6 and −4.4, making it the third brightest object in the sky after the Sun and Moon in apparent magnitude. To separate any of the dusky features on its bright disc, filters are a must.

Most of the difficulties inherent in getting a good telescopic view of Venus involves its proximity to the Sun. Venus is never more than 47° from the Sun in the sky. Also, for a large fraction of the time, it lingers in the solar glare and makes observation difficult. But the greatest problem for telescopic observers is viewing Venus when it is high enough to avoid the poor "seeing" we encounter when we look through a large proportion of Earth's atmosphere. To overcome this, Venus can be observed high in the sky in the middle of the day, and many observers feel that this is the best time to look at it. This is possibly due to the well-known visual effect that the "surface" brightness of Venus is very great, but so is that of the bright sky in the same field of view; thus the brightness tends to cancel out, giving the planet a lower contrast and thus rendering faint detail better. But finding even an object as brilliant as Venus in broad daylight is usually very difficult for those who do not know how to go about it and requires some work even for the veteran observer. Such an exercise is rendered easy with a telescope fitted with effective setting circles or a GOTO mount.

There is an additional problem with seeing detail on Venus. The planet is eternally shrouded in cloud, revealing very few features. This cloud, which is mostly made of carbon dioxide, sulfuric acid and water vapor, prevented all attempts to learn what conditions were like on this world until the advent of radar and satellite technology. Until then, Venus was considered Earth's twin, not only in size but in surface conditions, too, being compared to Earth's early geological periods such as the Carboniferous. In fact, many flights of fancy were penned about Venusian conditions, one of the more famous being Tennyson's quote from "Locksley Hall, Sixty Years After":

> Venus, smiling downward on this Earthlier Earth than ours,
> Closer on the Sun; perhaps a world of never fading flowers,
> Hesper; whom the poet called the bringer home of all good things,
> All good things move in Hesper, perfect peoples, perfect kings,
> Hesper – Venus – were we native to that splendour, or in Mars,
> We should see the globe we groan in, fairest of their evening stars.
> Could we dream of wars and carnage, craft and madness, lust and spite,
> Roaring London, raving Paris, in that peaceful point of light?
> Might we not glancing heavenward on a star so silver fair,
> Yearn and clasp the hands and murmur "would to God that we were there."

Given what we now know of the surface, few would like to visit the planet and wonder about our "fair Earth" from its perspective. The yellow–white cloud cover of Venus that we see through the telescope is a layer of sulfuric acid haze very high above the crushingly heavy atmosphere of the planet. The huge amount of carbon dioxide in the atmosphere produces a "greenhouse effect," creating surface temperatures of around 475 K. Adding to the planet's mystique, each day on Venus lasts slightly longer than the planet's year (the Venusian day is 247 Earth-days long, while the Venusian year is 224 Earth-days long), resulting in Venus spinning backward, or having retrograde axial motion. The origin of this strange rotation is one of our most basic questions still unanswered regarding the nature of this hellish world, although theories such as huge impacts and tidal distortion during the formation process are extant. Whatever history may record, Venus certainly is not an identical twin of Earth! Tennyson's romantic summation above provides an insight into the professional and public perception of the planet at the end of the nineteenth century.

From the observational point of view, Venusian apparitions fall into two classes: those in which Venus is the evening star at greatest eastern elongation and those in which Venus is the morning star and is thus at western elongation. The Greeks and several other early cultures considered these apparitions to be separate entities, with the result that Venus had two names—Phosphorus and Hesperus—which denoted these differing appearances.

Filters for Observing Venus

Low in the dusk after coming out from superior conjunction, observers can see Venus as a round little disc hardly more than 10 arc seconds across. Venus then goes through an almost full set of phases as seen from Earth, from almost full or "gibbous" after superior conjunction and then dwindling until it becomes "new" or invisible to us at inferior conjunction. During this time from full to new, the size of the Venusian disc increases to a 68″ maximum at slim crescent, and we can see at a glance through the telescope almost just where Venus is in its orbital journey in relation to Earth. It is recommended that amateur astronomers make some observations of Venus through every phase it displays if possible. All too many observers will wait until Venus nears its critical half-lit phase to start watching, and it is quite possible that some interesting sights are being missed.

One of the advantages of observing Venus, as mentioned earlier, is that it can be seen in the daylight. Once its position is known setting circles or GOTO mounts will find the planet which looks like a bright star in a low power eyepiece. Figure 4.3 reveals Venus as seen shining in a daytime sky. The violet No. 47 filter shades the vivid blue sky and allows the bright planet to shine out, but also adds contrast to some of the surface features—which in this case are nothing more than clouds. Markings on Venus are difficult to see, and many hours of patient observation and experience are required to see them well, but the added contrast of the violet filter does make a difference to the view.

Using a Wratten No. 15 deep yellow filter has been recommended by various observers, as the filter brings out irregular deformations of the terminator that are very striking around the time when Venus is half lit, near the time when it is at greatest elongation, which is coincidentally the time when it is highest in the sky at sunset or before sunrise. These deformations are of course due to the varying heights of the cloud cover in the atmosphere and can be seen through a modest telescope with high magnifications under good seeing conditions. Figure 4.4 demonstrates Venus seen through such a filter.

An intriguing facet of observation is that Venus ought to appear half lit exactly at greatest elongation, but usually it appears so at least several days before in the evening apparition. This dichotomy is known as the Schroter effect, after Johann Schroter, the man who first noticed over 200 years ago this lag between the phases of Venus and its elongation timings. To see the maximum contrast in the Venusian clouds at this time either a deep red Wratten No. 25 or even a violet No. 37 filter

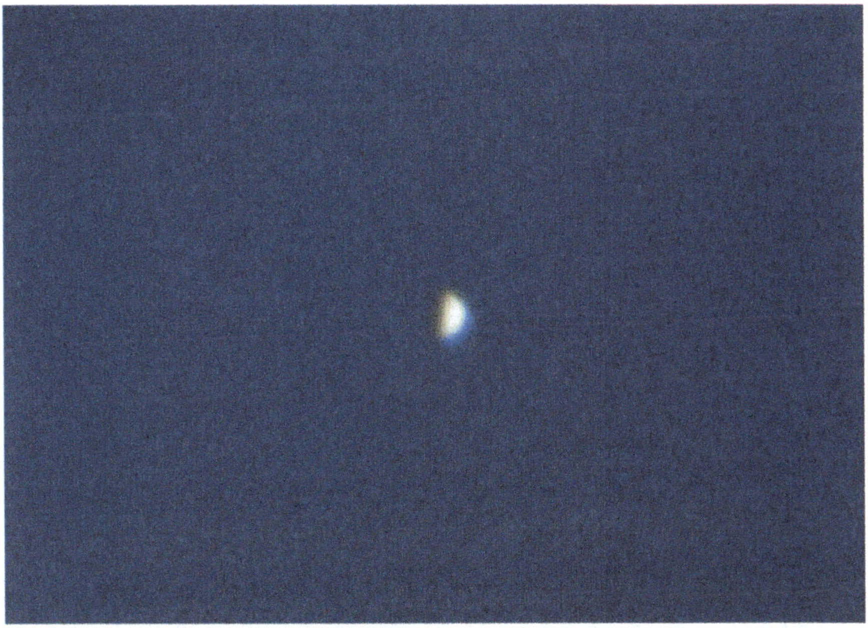

Fig. 4.3 Venus through a Wratten No. 47 filter

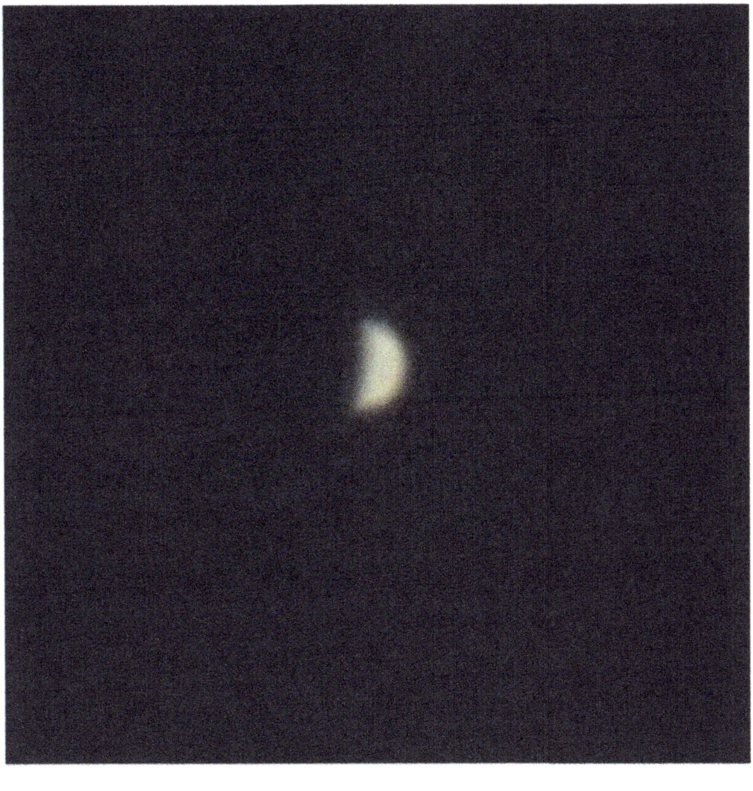

Fig. 4.4 Venus half phase, Wratten No. 15

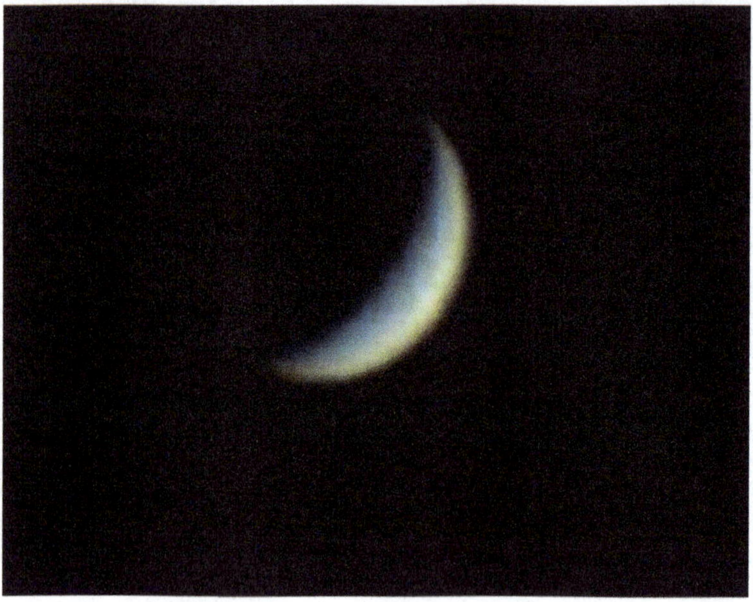

Fig. 4.5 Venus crescent and No. 80A filter

can be used, but again, very like Mercury, the observer has to be extremely wary that any contrasting features are real rather than imaginary.

However, observers of the past noticed that Venus had a 4-day atmospheric rotation, which was called into question when its axial rotation was found to be 247 days. The *Magellan* spacecraft however set the record straight in the early 1990s when it discovered that despite the long axial rotation, the upper atmosphere had a 4-day rotation period! Something in the cloud decks *had* been noticed with the assistance of a filter. Figure 4.5 shows the crescent planet in the twilight assisted by a Wratten No. 80A filter.

Try and see if you can spot any contrast features on the cloud tops of the planet. These will most probably be slight shadings in hue or its irregular terminator. Spacecraft have imaged a polar hood in the clouds of Venus that has a slightly different color to the clouds elsewhere so it may be possible to see this under the right conditions with the correct filter. Trial and error with Wratten blue, yellow, green and red filters may provide such opportunities.

A rare event that gives the observer a glimpse of the planet in full is that of a Venusian transit. This intriguing phenomenon has inspired and aided astronomers to make some crucial discoveries. Venus, being inside the orbit of Earth, presented historical opportunities to glean information on the size of the Sun and the distance to our nearest star. As the orbital element of the planet is known fairly accurately, the timing from when the planet first touches the Sun's disc to the time when it exits the disc enables the observer to calculate the solar diameter. Measurements of the transit from different locations will also give an accurate distance to the Sun.

Sadly, transits of Venus are generally very rare. Two occur separated by 8 years, but then there is a gap of either 121.5 or 105.5 years before the next pair! The last pair was in 1874 and 1882; the last pair of transits was on June 8, 2004, and June 6, 2012. The next pair will be in 2117 and 2125. The astronomical importance of a transit is now significantly diminished simply because the Earth–Sun distance can be more accurately achieved using radar, but such technology did not exist in the eighteenth century, when the need to find the Earth–Sun distance became a major political race between Britain, France, Spain, Holland and Italy. Sir Edmund Halley first suggested measuring the transits of the inner planets across the Sun's disc as a way of determining the distance to the Sun.

Halley was then the Astronomer Royal at Greenwich Observatory and involved in settling the problem of longitude. This, and determining the Earth–Sun distance, were probably the two largest problems in eighteenth-century astronomy. Needless to say, the race included some spectacular debacles. Jean-Baptiste Chappe d'Auteroche of the French Royal Academy of Sciences went to observe the 1761 transit from Russia, but was attacked by mobs who thought his experiment would somehow interfere with the Sun. His countryman Alexandre Guy Pingre was rained out when he attempted to observe the transit from Madagascar, and was subsequently captured by the British, but took refuge against his bad luck in liquor. Worse was to follow for Guillaume Le Gentil; attempting a trip to Pondicherry in India to observe the transit, he was becalmed in the Indian Ocean on the big day as British troops had routed the French garrison at Pondicherry. The French and British eventually made a concordat and returned the town to the French, and Le Gentil was able to build his observatory, but he had missed the transit.

Undaunted, he waited for the next transit 8 years later, only to have the Sun obscured by clouds on the day! Returning home in misery, he contracted dysentery, was shipwrecked, but eventually made it back to Paris 11 years after he had set out, only to find he had been declared dead and his estate looted! Few astronomers have had to put up with so much.

Figure 4.6 shows Venus in transit across the Sun in 2004. The 2012 event was widely photographed by legions of observers through solar filters ranging from Mylar and Baader film to Hα and Calcium II filters. This photograph was taken by the author using a Canon 350D SLR through an 80-mm telescope fitted with a Baader Astro film filter. Exposure was 1/350th of a second at f6.2 and ISO 800, and the image was then processed and colored in Photoshop™.

Hopefully, you will not have to suffer such rigors as the astronomers above if you go out and observe this fantastic planet and ponder its physical characteristics. Using filters in the Wratten system will enable you to see some of the depths of cloud in the upper atmosphere of the planet, but careful observation is required. Try testing various Wratten filters across the range from red to blue on the planet under moderate to high magnification and attempt to discern any features that can be followed over a period of a few hours if possible. As Venus hangs in the sky for several hours after sunset, it presents more of an opportunity for filter observation than its inferior counterpart Mercury.

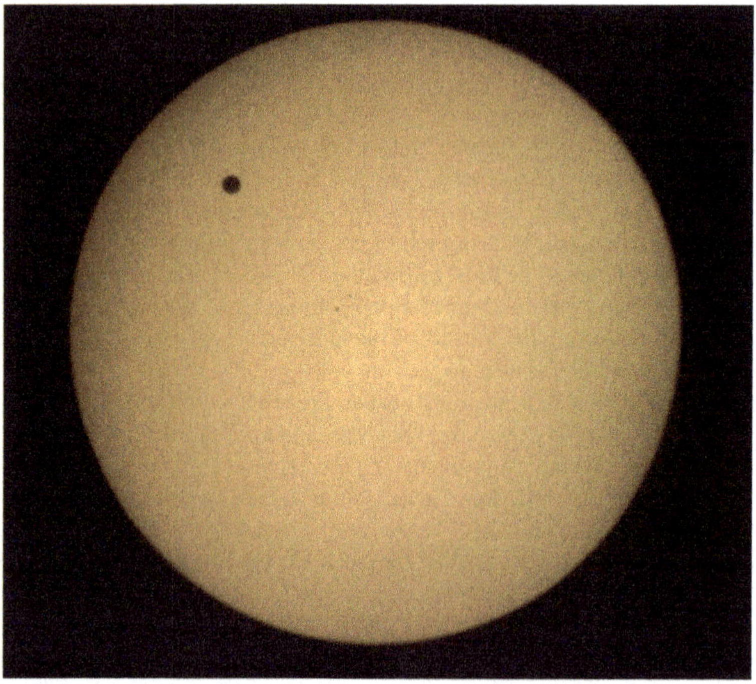

Fig. 4.6 Venus in transit, Baader paper filter

The table below provides a handy guide to the types of filters that will aid the visual observation of Venus.

Wratten No	Color	Features
No. 15	Deep yellow	Contrast features, possible cloud layers, separation of sky
No. 25	Deep red	Contrast features and darkening of sky, clouds.
No. 80A	Blue	Contrast features, possible cloud layers, darkening of sky
No. 47	Violet	Contrast features and darkening of sky

Mars: The Red Planet

The most fascinating world beyond our own is the planet Mars. No other planet holds such potential for human exploration in the next 25–50 years. Mars appears to be Earthlike enough to make its colonization seem almost inevitable. Mars draws, demands our attention. Of all the planets of our Solar System not one has the stature of Mars in science, myth or fantasy. Mars is a magical name, heavy with allusions, aspirations and grand dreams unfulfilled; it is a repository of creativity and desire from which spring new insights, innovations and technologies.

For naked-eye observers, Mars is also the planet with the most distinctive color—a deep orange so impressive that it has inspired most cultures to call it "the Red Planet." For telescopic observers, Mars is the only terrestrial planet that shows real surface features, and its apparition can last an appreciable time, enabling observers to familiarize themselves with the surface.

More than any other planet, Mars is also an enigma; surface conditions are never really good for any length of time, and, even when visible, it appears tiny and rather bland through the telescope. In fact, the periods when it swells large enough to have a wealth of details visible on its disc occur only at infrequent intervals and only at certain of its oppositions. It is therefore easier to discuss an average apparition—one in which it comes neither extremely close to us nor misses us by a great distance.

Mars orbits at about 1.5 times as far as Earth from the Sun, taking almost twice as long as Earth to complete an orbit. It requires about 780 days for Earth to catch up to Mars a second time, or about 780 days between one opposition of Mars and the next. This synodic period is far larger than that of any other planet and determines the way Mars tends to be seen through telescopes—distant, small and disappointing 1 year, close and relatively large the next. An apparition of Mars lasts far longer than that of any other planet, as does its period of invisibility near the Sun between apparitions, thus adding to the mystery surrounding the planet and the frustration of observers.

This legendary propensity to be awkward has left its mark on the human psyche. Fantastic tales have entertained humankind since the dawn of civilization, but an avid and ready audience has always been available for any story that features the planet Mars. There are not many in western culture who have not heard of H. G. Wells and his famous *War of the Worlds*, a novel that introduced the warlike Martians to the public consciousness. Science fiction stories have featured regular references to Mars as a place likely to be inhabited by alien life, or at least by ourselves 1 day. Such tales drew upon the then current astronomical theories of Lowell, Schiaparelli and Flammarion, which gave such stories verisimilitude, and, interestingly, drove the development of some modern technologies.

The young American scientist, Robert Hutchings Goddard, was entranced by the possibility of spaceflight to or from Mars. Reading a serialization of Wells' *War of the Worlds*, he began to ponder the existence of life on other worlds and how to contact such beings. By 1926 he had launched the first liquid-fueled rocket and developed the theories that would propel humans out of their Earthly home. Similarly, the SF influence of Kurd Lasswitz's *On Two Planets* (1897) motivated such rocket pioneers as Hermann Oberth, Arthur Rudolph and Werner von Braun. In Russia, imaginative rocket fiction from the pen of Konstantin Tsiolkovsky evolved into reality by the late 1930s.

The work of these rocket pioneers was paralleled by science fiction writers such as Edgar Rice Burroughs, who wrote a series of books, starting with *A Princess of Mars*. Burroughs based his ideas of the Red Planet on the lurid "canal" theories of the astronomer Percival Lowell, creating a world that was dying as the planet dried up and the canals emptied, ultimately to be fought over by the green-, yellow-, red- and black-skinned inhabitants. Burroughs stories inspired not only a host of imitators but a generation of scientists and technicians who would later become involved in robotic exploration of the surface. Mars in popular culture was given an additional boost by the pulp magazines of the late 1920–1940s, wherein dramatic tales of survival, warfare and intrigue were played out on or over the Martian landscape. Stanley Weinbaum's seminal *Martian Odyssey* in 1935 cemented the human relationship to Mars by creating a fascinating, ecologically rigorous world populated by totally un-human, yet intelligent, Martians (tweels), while radio plays and cinema shorts such as Flash Gordon kept the possibility of Martian life (good or evil) in the public eye.

It is little wonder that the public is so fascinated with the Red Planet, yet at the same time it can be disappointed by the usual telescopic views! For the observer equipped with a moderately large telescope, observing Mars is almost as fascinating as its cultural and literary connections. As with any superior planet, an apparition of Mars begins with its first sighting low in the morning twilight after conjunction with the Sun. The reason why the sighting occurs many weeks after conjunction, compared to just a few weeks for Jupiter and Saturn, is twofold. First of all, Mars on the far side of its orbit from us appears decidedly dimmer than Saturn and much less luminous than Jupiter. It is a small planet and needs to come closer to Earth before any appreciable details can be seen. To be appreciated, it also requires a suitable elevation away from the glow of the bright twilight. Additionally, delays in the apparition sighting are due to the eastward orbital speed of Mars, which is so much greater than that of the other superior planets. It takes a while to catch up, which results in oppositions at approximately 2-year periods.

When Mars does begin to get reasonably high in a dark sky before dawn, notice how it keeps almost the same altitude week after week while the constellations are carried far more rapidly westward by Earth's own (eastward) motion in orbit around the Sun. The brightness of Mars is only about second magnitude at this stage—quite dim—and consequently many a novice may have trouble spotting it among what may be as many as a dozen or so stars as bright or brighter in this general section of the heavens.

Because Mars is so much closer to Earth than Jupiter or Saturn, its retrograde loop against the starry background as Earth begins to overtake it is larger than any of the other superior planets, yet lasts for a shorter time. When you see the Red Planet halt and start to go westward with respect to the background stars, you know that opposition is only about a month away. Observing the planet regularly will enable you to determine the retrograde motion and the likely opposition times.

When you set your telescope on the correct second-magnitude object and see a planetary disk, you may be surprised to discover it is tiny, only about 4 arc seconds in width. Most are disappointed with their first telescopic view of the planet, and even at a good opposition, the disc is still small and shows few details. Observers immediately notice the ochre color of the disk, but at this stage there are no visible features in small telescopes. When Mars is about 60° from the Sun, its disk has

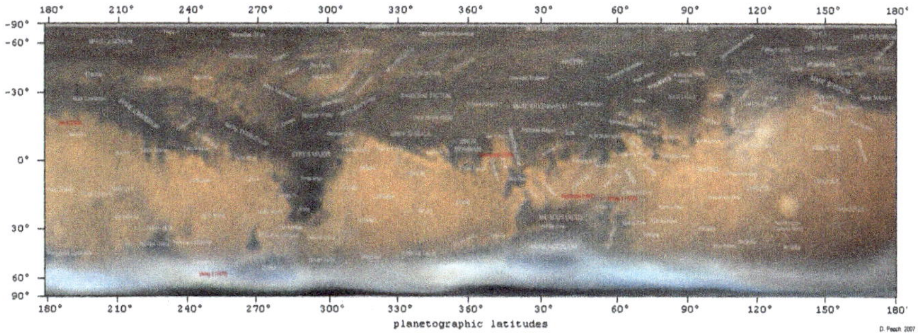

Fig. 4.7 Map of Martian (*Credit*: NASA)

grown to about 10 s, the minimum size at which useful observations of the planet may be possible. At this point observers can discover whether their telescopes and levels of experience can lead to seeing any features on Mars at this stage, even on nights of good seeing.

An average opposition of Mars has the planet rivaling or surpassing the brightness of the brightest star and swelling to about 25 arc seconds in width, or about five times Mars's apparent width when it is near conjunction. This is just about the threshold at which the few details on the planet begin to proliferate into numerous features in medium-sized telescopes. One of the best oppositions was the recent apparition in 2003 where Mars was the closest it had been for 60,000 years! It will never look the same at any opposition, however, because of the inevitable seasonal changes that have enticed so many astronomers to spend a lifetime observing this planet of mystery, speculation and awe. Included here is a map of the general surface features of Mars so that the observer can make comparisons to the eyepiece image. There are several applications for smartphones which also show detailed maps of Mars in real time and this author would recommend them. They can be downloaded, generally free, from any Apple or android online store (Fig. 4.7).

Mars reaches quadrature at 90° west of the Sun. It is then brightest around dawn or early evening and shows a remarkably wide shadowed edge from *phase effect.* By this stage, Mars has improved to roughly 0 magnitude and is on average about 10 arc seconds across. Medium-sized telescopes in the 200 mm+ range now begin to show a few features to the careful observer under good seeing conditions. A few of Mars's darker markings can be spotted. The most prominent is often the triangular Syrtis Major, located not far north of the Martian equator. It is a difficult feature to miss at this stage if it is near the central meridian of Mars, facing Earth, at the hour you are looking. A white glint of a polar icecap is also likely to be spotted, as are other hazy features of light and dark variation across the surface.

The appearance of even these most prominent of Martian features is greatly dependent on the current seasonal conditions on Mars. Some of the seasonal effects, such as the melting of polar icecaps, can be predicted fairly well in general. Others, like the famous dust storms, are more irregular in occurrence and varying in scope so that no one can be sure of when or even whether they will appear. Once in a while, a planet-wide dust storm can hide all the Martian features for many weeks. Thus, you can never tell just what you will see when you gaze through a telescope at this surprising world. Considering its variable pattern of storms, seasonal changes and polar cap growth and shrinkage, it is unsurprising that many thought Mars a likely abode for life.

Observing Mars with Filters

Probably the best filter to use on Mars is the Wratten No. 25 red, which gives maximum contrast to almost all of its surface features. Obviously with such a deep colored filter a telescope of 150 mm and above will be the most effective, as the planet will also demand high magnification to discern any surface features at all. With the No. 25 filter, the areas such as the Planitia deserts and darker areas such as Syrtis Major become more defined, and the filter also separates out the polar caps from the overall ochre hue of the planet.

However, the planet's glare and general fuzziness due to Earth's turbulent atmosphere, and the atmosphere of Mars combined with its occasional low elevation in our skies, render the image less than spectacular. During such times the filter above can be substituted for the Wratten No. 21 orange filter, which does make a great difference by reducing contrast, blocking blue light and allowing fainter detail to be glimpsed as can be seen in Fig. 4.8 here.

Many observers have also noticed good contrast with the use of the Wratten No. 15 deep yellow filter, which darkens areas of less reflectivity on the Martian surface and thus make large areas of the southern hemisphere and northern areas such as Acidalia stand out. The No. 58 green filter has also been used by many Martian observers to darken red and blue features, and it has been recorded that use of this filter also enhances frost patches, surface fogs, and features in areas close to the poles of the planet.

Mars has enormous seasonal variations across its disc. Observers of the past noticed a wave of "darkening" that crossed the Martian surface during the onset of northern or southern summers. Mars's axial tilt is very similar to Earth's at 23°, so these differences in appearance were interpreted to be the growth of vegetation. The orange No. 21 filter will bring out the darker features associated with this phenomenon, but close and regular inspection of the planet is recommended. Some observers have noted a distinct difference during such times of summer onset by using a yellow or light green filter in the Wratten Nos. 8–11 range, but it is recommended that such filters be used only once some experience of observing Mars has been gained. Others have noticed that the darker areas under such phenomena are rendered more visible and separated from the planet background if the deep red Wratten No. 25 is used.

As Mars has a thin atmosphere, occasional clouds can be seen near the large volcanic areas of the equator such as Olympus Mons and the Valles Marineris.

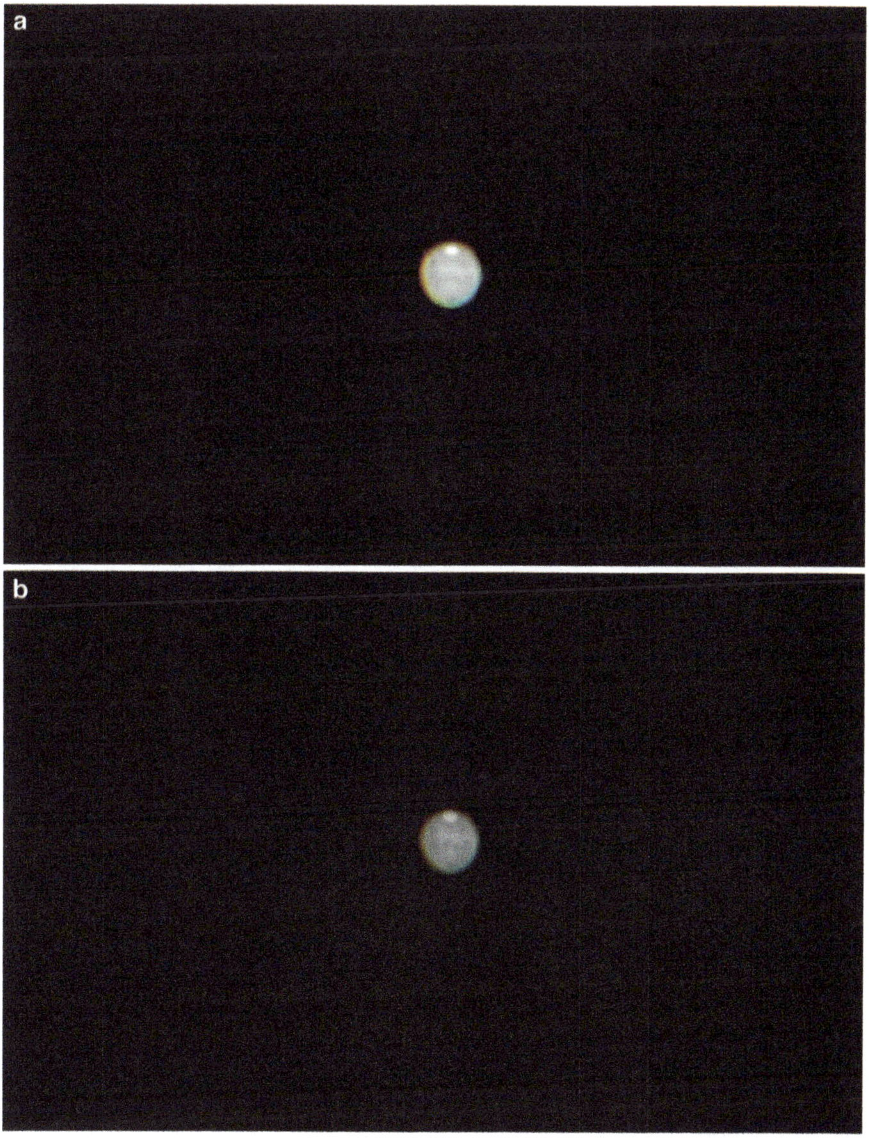

Fig. 4.8 Unfiltered (**a**) and filtered (**b**) Mars

Observers recommend the use of the blue Wratten No. 80A at such times to increase the contrast between cloud features and the surface, though obviously sharp eyes and some experience would be needed to discern such details. Nevertheless use a wide range of filters to examine the planet, as one never knows what could become visible. Using Wratten filters on a multi-faceted gem of a planet such as Mars may enhance features that were previously unseen by ordinary visual observation.

It is also worth considering that such filters are appropriate for CCD or webcam imaging of the planet. Some observers imaging with DMK cameras recommend the use of the Wratten No. 21 orange filter to increase the contrast in light or dark areas and to bring out any polar features such as icecaps, etc. Nonetheless, the table below gives a ready guide to the types of filters to use while observing Mars.

Wratten No	Color	Features
No. 25	Deep red	Contrast in surface features, enhances fine surface details, dust cloud boundaries, and polar cap boundaries.
No. 21	Orange	Increased contrast between light and dark features, penetrates hazes and most clouds, and limited detection of dust clouds.
No. 58	Green	Darkens red and blue features; enhances frost patches, surface fogs and clouds
No. 80A	Blue	Shows atmospheric clouds, limb hazes, equatorial cloud layers, polar cloud hoods, and darkens reddish features
No. 47	Violet	Reveals clouds, limb haze and darkens reddish features

Jupiter: King of the Planets

Jupiter is not only the largest of the planets in true size, but it is also the one whose combination of size and distance usually makes it appear largest as seen from Earth. On the infrequent occasions when Venus gets close enough to loom larger than Jupiter, only a small fraction of Venus's disk is lit. Jupiter is always a full on disc and is simply unmistakable and a wonderful sight even in the smallest telescope. Even a pair of binoculars, steadily held, will show the starry spots of the Galilean moons. Jupiter is also a very bright planet, second only to Venus from our Earthly viewpoint and can reach up to magnitude −2.6 and so again adds difficulties in contrast to the eye between the bright disc and the dark sky background. Filters therefore become extremely helpful in reducing the planet's glare and increasing the contrast between features on its surface.

Firstly, it is advisable to become acquainted with the largest features one can observe on the planet—or, rather, in the planet's clouds, for the only parts of Jupiter we ever see are its outer cloudy layers. The diagram that follows later will be invaluable in aiding you to observe the planet, even though at first glance you may not be able to see the underlying complexity shown in the diagram.

The most prominent features on Jupiter to see and record are the dark bands running parallel across Jupiter's bright face. Even a small telescope of barely passable quality can show a few of them. When Galileo looked at the planet in 1610, he recognized that the Jovian system was different; an observation that heralded the beginning of our modern scientific age. Galileo's telescope was a mere 1 in. in diameter, yet produced a revolution unlike nothing prior; so, too, may your perceptions of this system be overturned by the view through a modest 'scope.

The Jovian system has attracted planetary observers in droves, as there is such a wealth of spectacular and changing detail on the cloudtops. Jupiter has figured largely in astrology as the king of the planets, an appellation that has attracted artists and composers, Mozart and Holst dedicating a symphony, or parts of, to the planet. In fiction Jupiter has been the focus of several famous works by Arthur C. Clarke. His 1972 short story "A Meeting with Medusa" introduced the idea of sentient life forms on gas giant planets, while his seminal *2001, A Space Odyssey* has all the action of humanity's first meeting with extraterrestrial intelligence take place in the vicinity of the Jupiter system. In recent years, intense speculation has abounded in astrobiology circles as to the likelihood of life emerging on the moon Europa, which is now known to harbors a satellite-wide ocean beneath its icy crust—another idea pioneered by Clarke in *2010 Odyssey Two.*

The Galilean moons are a very prominent feature. Outward from the planet they are: Io, Europa, Ganymede and Callisto. However, as they revolve around the planet, they do not maintain this outward configuration, and it is necessary to examine a monthly astronomy magazine or software database every day to get their positions. Occasionally, they transit the planet, their shadows produce dark eclipse shadows on the cloud belts or they disappear as they are occulted by Jupiter. The dance of these satellites can become a source of endless fascination, and the whole system is an obvious target for budding astrophotographers.

After the moons, the planet's most obvious features are the dark bands that cross it. These bands are called *belts.* The lighter stripes between them are known as *zones.* The belts and zones are cloud features that are stretched out into planet-encircling parallel bands by Jupiter's rotation, which is the most rapid of any Solar System planet's rotation (less than 10 h per single rotation). A look at the belts and zones thus immediately tells you which way the rotational axis of Jupiter is aligned (perpendicular to the belts and zones) and therefore where the polar regions are located. For ease of identification the following map of the Jovian belt and zone system may be useful (Fig. 4.9).

You can also tell where the poles are and where the equator of Jupiter is by another property of the planet that is quite noticeable in small telescopes—its shape. Jupiter is decidedly *oblate.* This means its diameter measured from pole to pole is less than its equatorial diameter. Jupiter's equatorial diameter is about 6 % larger than its polar diameter. It is relatively easy to spot this feature.

The most prominent of Jupiter's belts are usually the South Equatorial Belt and North Equatorial Belt, located to either side of the planet's equator. They appear as a brownish or reddish series of lines bounding the whiter equatorial zone and are very prominent even in small 'scopes. Seeing such detail requires a little patience and application rather than a quick glance through the eyepiece. Numerous belts and zones, as well as many other features, may be seen by a good observer with a medium-sized or large telescope. Beginners should content themselves with first identifying the prominent belts and zones and watching the dance of the satellites as they wheel around the planet.

Novices can also note what colors they glimpse in the belts and zones. These may at first seem merely gray and yellow–white, respectively. But further

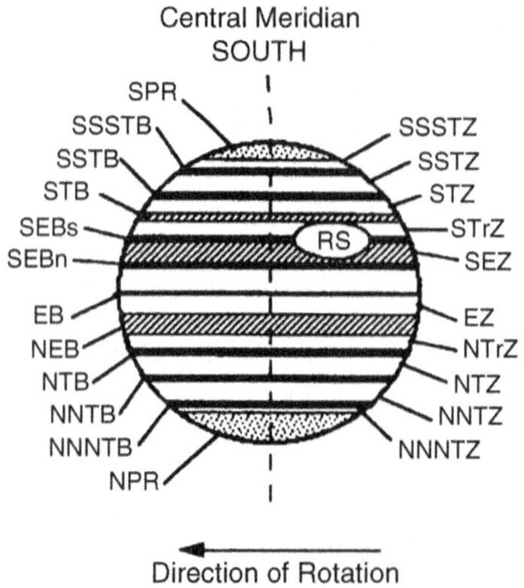

INDEX TO ABBREVIATIONS

ZONES
EZ - Equatorial Zone
SEZ - S. Equatorial Zone
STrZ - S.Tropical Zone
STZ - S.Temperate Zone
SSTZ - S.S.Temp. Zone
SSSTZ - S.S.S.Temp. Zone
NTrZ - N. Tropical Zone
NTZ - N. Temperate Zone
NNTZ - N.N. Temperate Zone
NNNTZ - N.N.N. Temp. Zone

BELTS
EB - Equatorial Belt
SEBn -S. Equatorial Belt North
SEBs -S. Equatorial Belt South
STB -S. Temperate Belt
SSTB -S.S. Temperate Belt
SSSTB -S.S.S. Temperate Belt
NEB - N. Equatorial Belt
NTB - N. Temperate Belt
NNTB - N.N. Temperate Belt
NNTB - N.N.N. Temp. Belt

OTHER
SPR - S. Polar Region
NPR - N. Polar Region
RS - Great Red Spot

Fig. 4.9 Map of Jupiter (*Credit*: British Astronomical Association). Photo courtesy of Graham Hurley

examination (or a larger telescope) should reveal at least a hint of brown (you might even say a slight ruddiness) in the belts and distinctive yellow hues in the zones. Remember, however, that even on nights of good seeing color estimates should really be tried only when the planet is fairly high in the sky.

On the subject of color, the famous Great Red Spot (GRS) of Jupiter is often very pale in color and often difficult to see, belying its epithet. It is located near the edge of the South Equatorial Belt but can be very difficult to spot in small instruments. Most of the time the GRS hollow is visible as a "bite" in the cloud belts rather than the spot itself. The whole feature is visible with some dedication and patience even through a modest instrument. The planet is an endless source of wonder and its ever-changing configuration should be viewed whenever it is above the horizon.

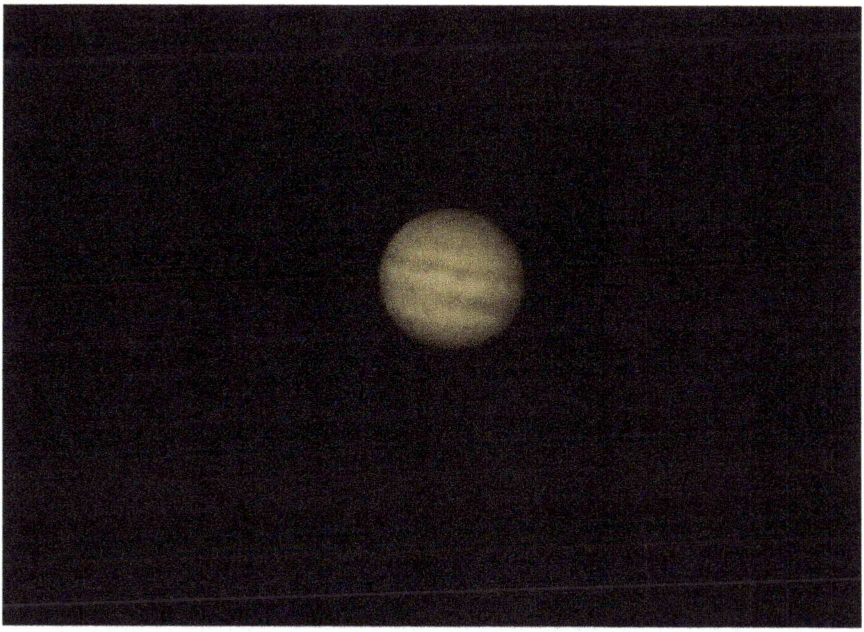

Fig. 4.10 Jupiter using a Wratten No. 21 Orange filter

Filters for Observing Jupiter

Jupiter is an amazingly detailed object, and as its size varies from 30 to 52 arc seconds during its apparitions it subtends the largest body of any of the superior planets. As noted earlier, Jupiter is also a very bright planet. Using nothing but the naked eye, most observers can gain enough detail to satisfy their requirements, but many astronomers also find that blinking the orange No. 21 filter into the eyepiece does render the planet literally in a different light. Subtle shading and detail can be seen flashing into view as the wide bandpass enhances the reddish areas such as the Great Red Spot on the planet and the equatorial and temperate belts as can be seen from Fig. 4.10 here. How much of this detail can be seen by a keen eyed observer without a filter? The filter also sharpens the boundaries between the belts and zones and increases the contrast in the bright zones by lowering their light transmission, enabling some faint details to be seen under higher powers.

A full range of filters can be used on Jupiter, though the color correction filters do not often give the same blaze of details that the blocking filters do. For example, the effect of the No. 38A filter can be seen in Fig. 4.11, where the delicate cloud bands of Jupiter show up in marked contrast to a very glare-filled view through an unfiltered eyepiece. The red banding is a little more enhanced, but so are the white

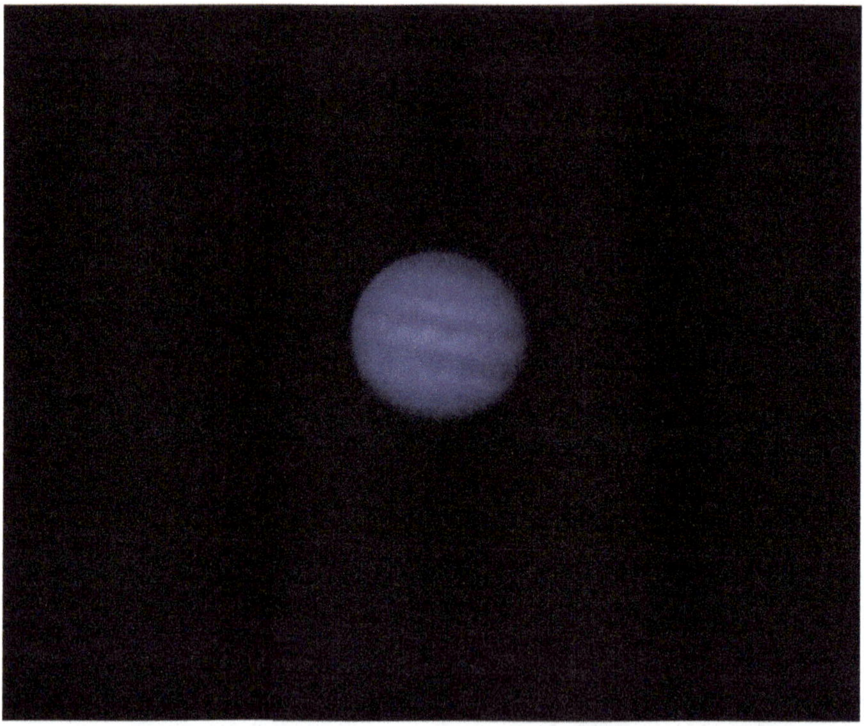

Fig. 4.11 Jupiter Wratten No. 38 dark blue. Photo courtesy of Graham Hurley

zones between the belts. It is easy to see how the areas of shear between the belts and zones and the melding between them reveal more subtle detail than in the orange No. 21 filter, as seen previously. Use of the dark blue Wratten No. 38 filter really does work, and the substantial reduction in glare allows the astronomer to work without tiring the eye, and since the planet is large, high magnifications even on relatively small telescopes can be used.

In fact, the filter that does all of the above in enhancing the belts and zones and revealing marvelous detail in the cloud tops under steady atmospheric seeing is the Wratten No. 58 green filter. Although this filter colors the planet in a rather surreal hue, the detail visible in the separation areas between bright zones and the darker belts is fantastic. Shear areas are brought out in sharp relief, and the reddish areas of the belts become broken into a maze of detail under high magnification, with white high clouds floating above the main belts and the limb of the planet easily visible as broken into an irregular sawtooth pattern. The belts and zones represent areas of different pressure and height in the cloud layers of Jupiter. The GRS and its hollow become far more visible and the poles themselves a little more detailed.

Some observers notice that the use of the Wratten No. 15 yellow filter deepens the detail noticeable on festoons of cloud close to the poles, while the Wratten

No. 25 deep red filter enhances the view of several of the giant oval storms visible close to the equatorial belts. It is thus obvious that the full range of longpass filters are appropriate to use on Jupiter, bringing out the wealth of detail on this giant planet. Below is a quick guide to the filters and features that are generally visible with their use.

Wratten No	Color	Features
No. 38	Dark blue	Enhances the contrast within the bright zones and sharpens boundaries of faint cloud currents
No. 21	Orange	Contrast sharpened between belts and zone boundaries
No. 58	Green	Increased contrast in belts and zones and polar areas better defined. More detail enhanced in belts
No. 15	Yellow	Contrast sharpened between belts and zone boundaries
No. 25	Deep red	Increased contrast in bright zones and better resolution of white oval storms

Saturn: The Ringed Planet

The two most striking sights in a telescope for most first-time viewers are the Moon and the planet Saturn. With good seeing and a properly collimated telescope (even a rather small one), the body and rings of Saturn are beheld with a sharpness that photos cannot convey. The impression one is left with when viewing the planet for the first time are varied, but inevitably, most people go to the front of the 'scope in order to look for the photograph that the astronomer has placed there to fool them! Such is the glory of Saturn through the telescope. Even veteran observers never grow quite tired of this basic sight, even if the presentation angle of the rings never changed. Nevertheless, dedicated observers find an interest beyond the aesthetic enjoyment of the study in illuminated geometry or statuary of light, shadow, and shapes that is Saturn.

As a superior planet, Saturn behaves in the sky much like Jupiter—first dawn appearance after conjunction with the Sun, west quadrature, start of retrograde motion, opposition (rising at sunset), retrograde motion, east quadrature, and last dusk appearance just before conjunction with the Sun. But Saturn is far slower rotating than Jupiter—so slow that we can find it practical to talk about where Saturn will be and what Saturn will be doing over the course of not just a year but a whole decade. Saturn takes 29 years to go around the Sun and thus can be found in the same constellation for 2 consecutive years before moving on.

This slow motion across the sky led ancient people to associate the planet with the onset of old age—a fact rendered artistically by Holst in his *Planets* suite. This slow movement was also held to be responsible for slow yet unstoppable growth of any inhabitants, as imagined by Voltaire in his satirical *Micromegas*, detailing the visit of Sirian and Saturnian giants to Earth.

The unlikelihood that a gas giant could be inhabited made little difference to the literary or scientific establishment of the eighteenth century. At that time people did not know the composition of the planets, and it was therefore in keeping with then current speculation that such worlds were like Earth. No less an astronomer than William Herschel considered all the planets, including Saturn, to be inhabited.

The color of the planet is quite striking; it almost always appears as a steady gold point of light of between about magnitude 0 and 1, which is comparable in brightness with a number of the brighter stars; indeed at first glance it can be confused with one of the stellar beacons of the night, its baleful hue being alluded to by several ancient poets, including Homer.

A look through the telescope confirms the identity of the planet in spectacular fashion. Even binoculars or finderscopes can show that this image is elongated. A very small telescope and ×30 or ×40 magnification shows the rings as being tiny but definite. A slightly larger telescope and more magnification reveal that these are in fact rings in the plural. You may not see the thin black line that separates the two most prominent rings, the A and B rings, but the greater brightness and lightness of the inner (B ring) of the two make it stand out.

What are the rings of Saturn made of? Galileo's telescope was so poor that he thought they looked like "ears." About 50 years later, in 1655, Christiaan Huygens was the first to see that the rings encircle the planet without touching. But it was a matter of speculation as to what the rings were constructed from. Were they solid or were they gaseous or icy? The matter was eventually settled by the Scottish physicist James Clarke Maxwell in the mid-nineteenth century when he found by simple Newtonian physics that the rings must be composed of individual particles—unimaginably large numbers of them—orbiting Saturn at whatever speed is appropriate for a given particle's distance from the planet following Kepler's laws of planetary motion, though on a small scale. Scientific speculation as to their origin abounds, though there is little doubt that these particles are bits of ice and rock, probably the kinds of materials that aggregated to construct the planet and its moons in the early Solar System but became part of an extended equatorial system due to high angular momentum.

The rings are fascinating. One of the most intriguing features of observing Saturn is the ever changing angle of its rings. As the planet is tilted on its axis, the rings present themselves at various degrees. For several years the northern aspect will be visible, and then the rings will thin to an almost invisible line before the tilt brings the southern aspect of the rings into view. The telescopic view of the rings varies over a long time, although some of the changes leading up to the edge-on appearance of the rings happen in a matter of months or even weeks or days.

What basically happens during this phenomenon is that the angle from which we view the rings changes. The rings stay tilted at the same angle with respect to Saturn and its orbit, but Earth's vantage point shifts to give us anything from about a 27° to a 0° angle of the rings with respect to the horizontal. In other words, the rings can be fairly well tilted ("open"), showing us a good view across the vast expanse of their top or bottom (north face or south face), or the rings can be presented to us perfectly edge-on or "edgewise" from a sideways view. The marvelous

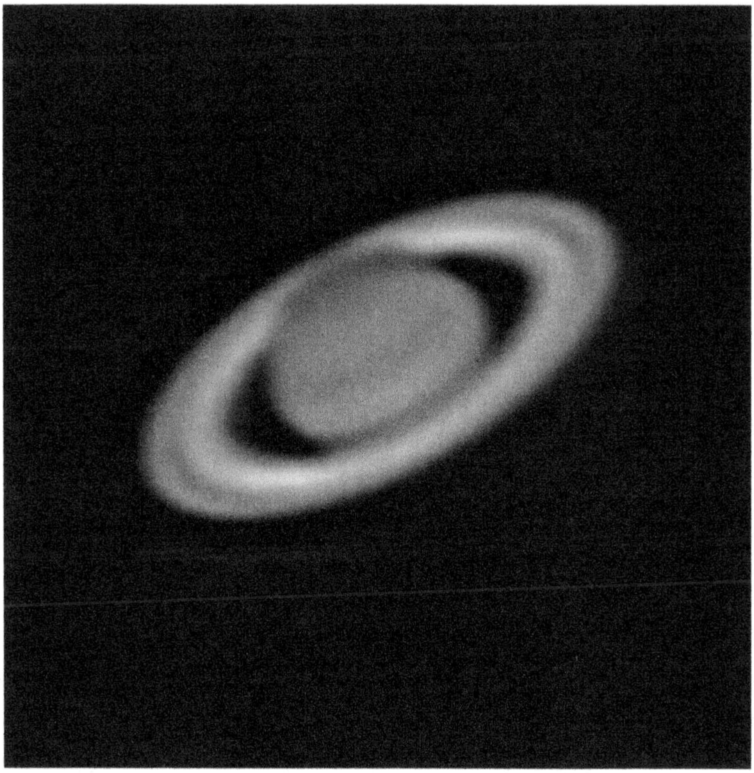

Fig. 4.12 Saturn ND filter

thing is that, although Saturn's two most prominent rings span across a distance of about 300,000 km in the other two dimensions, their thickness is as little as 50 km. The planet and its ring system can be seen here in Fig. 4.12, as seen through a neutral density filter, and its majesty and wonder are visible at a glance.

Although its superb ring system usually steals the show, the globe of the planet can be very interesting, too. As with the other gas giant planets, with Saturn we look at only the planet's clouded atmosphere. The *Voyager* and *Cassini* spacecraft passing Saturn found that it is not merely the planet's distance that makes its cloud features appear far fewer and subtler than those of Jupiter. The features really *are* far fewer and subtler due to on overlying layer of haze. There is also less activity out at Saturn; the sunlight here is less than 1 % of that impinging on our atmosphere, so solar heating plays a smaller part in the weather features of the planet. However, Saturn has an internal heat source that makes its weather systems as dynamic as Jupiter's. Fortunately, there are times when features rise up through the haze so that some rare but marvelous detail is sometimes glimpsed by veteran observers, such as the huge white storms that happen every 30 years or so.

Another aspect of Saturn's globe that is always visible, but mostly overlooked, is its oblateness. This is not surprising when we consider that Saturn is the only planet with an average density much less than that of water; and that it spins almost as rapidly as Jupiter. Saturn's diameter through its poles is about 10 % less than through its equator—roughly one entire Earth-diameter different! The reason that even novices do not notice the oblateness more often can be blamed on the rings. Besides distracting us from evaluation of the planet's shape, perhaps they also lead us to subconsciously think that our perception of the planet's oblateness is an optical illusion induced by the influence of seeing the rings elongated in the planet's equatorial plane. In any case, when the rings are near edgewise and almost vanished, a look at Saturn's globe immediately shows how significantly oblate it really is.

The satellites of Saturn are very varied. Usually visible as an eighth magnitude starry point is the giant moon Titan, shrouded in a methane and nitrogen haze and containing the possibility of pre-biotic chemistry on its surface, a feature that was investigated by the *Huygen's* probe in 2004. Fainter satellites such as Tethys, Enceladus and Dione are available to larger instruments, while Iapetus is of note due to its variable magnitude, one half of the moon being covered with a dark layer of compounds, possibly the result of a cosmic collision or outgassing from its surface.

Even if you cannot make out any of the belts or zones of Saturn on a given night, however, there are several basic sights to look for. Two are the shadow of the planet on the rings and the shadow of the rings on the planet. These sights are best visible around the time of the quadratures, for, although Saturn itself is too similarly distant from both Earth and the Sun to really show *a* phase effect, the rings are large and elongated enough to make it possible to see some of the planet's shadow on them and some of their shadow on it.

Filters for Observing Saturn

In the same manner as Jupiter, the planet Saturn responds well to the longpass filters that block light of particular wavelengths. However, due to the atmospheric layers of aerosol haze, features on Saturn are much more subtle, and require observers to concentrate their efforts in times of good seeing. Additionally, Saturn is smaller than Jupiter and is almost twice as far away from Earth as the king of the planets so its disc will require high magnification and telescopes in the 150 mm+ range to bring out the best of the details.

Once again the Wratten No. 58 green filter gives the best rendering of surface detail on the planet, as it differentiates the belts and zones quite well and reveals some features on the rings, too. The filter does not provide the wealth of details we saw on Jupiter, as Saturn has belts and zones that are smaller and smoother than its giant counterpart, yet they can be brought out from the overall yellow color of the planet nonetheless.

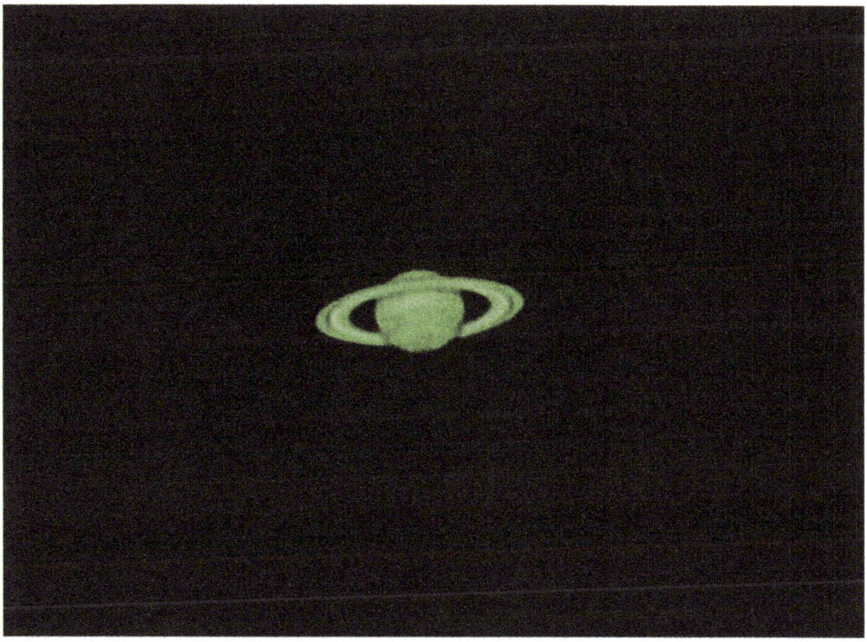

Fig. 4.13 Saturn using the No. 15 yellow filter

With the No. 58 filter the belts are noticeable as evenly distributed, thinner, with less range than Jupiter's and with all of the belts almost a uniform hue. Some observers report seeing some definition in the zone areas of the clouds with a Wratten No. 15 yellow filter, and some of this definition can be seen here in Fig. 4.13, which was taken with a Canon DSLR mounted on a 100-mm telescope at f10 and an exposure of 1/250 s at an ISO of 800, coupled to the No. 15 Wratten filter and processed in Photoshop™.

Visual observation of the planet reveals that the brighter zones appear off-white or even slate-gray or yellowish at times. However, in contrast, Saturn's belts exhibit bluish-gray, brown and reddish colors easily seen using the same filters as for Jupiter. Brighter patches sometimes appear on this Ringed Planet and are best seen by using the Wratten No. 11 yellow green or the No. 58, as detailed above.

The rings are a slightly different matter to the belts and zones of the planet. Many observers recommend the use of the Wratten No. 11 yellow green filter to bring out the details in the rings and enhance the Cassini division between the A and B rings whilst others have recommended the use of the No. 47 Violet filter to bring out subtle features in the rings and to see the Enke division in the A ring in better detail. This filter has also been claimed to render the belts and zones in greater detail on the surface of the planet too. Still other observers report seeing details to good effect by using the red Wratten No. 25 filter, so it seems obvious that these longpass filters are very versatile, and the detail they can reveal depends on our atmospherics, visual acuity, and the phase of Saturn and its ring system angle.

Saturn is a planet of unique beauty that continues to draw the attention of observer after observer. To gain the most from your observing, below is a table for quick reference for filtered observation of the ringed planet.

Wratten No	Color	Features
No. 11	Yellow green	Contrast on the rings and belt zones. The Cassini division
No. 21	Orange	Increased contrast in some belt and zone areas
No. 15	Yellow	Zone definition and some separation in the rings
No. 25	Red	Zone and cloud features, penetration of some haze layers
No. 47	Violet	Ring details and some enhanced features in the clouds
No. 58	Green	Bright cloud patches and storms, some enhanced detail in the rings

The Outer Darkness: Uranus and Neptune

It seems incredible to us living in the twenty-first century that ancient humans knew nothing about the existence of the three planets beyond Saturn. Pondering this, we can acknowledge the surprise of astronomers who discovered these worlds floating in the darkness at the edge of the Solar System. The story of their discovery is a tale of scientific insight, buffoonery and observational skill that highlights the nature of any human endeavor.

Uranus was discovered by William Herschel on March 31, 1781. Herschel was observing a bright disc-like object in the constellation of Gemini from the garden of his home in Bath, England. After watching it for several hours and noticing a slight movement against the background stars he recorded his observation and sent it to the Royal Society, believing the object was a comet. Calculations suggested that the object was in fact in an almost circular orbit around the Sun with a period of 84 years. The body was not a comet; rather it was a new planet, the first to be discovered in the age of science and one that doubled the size of the Solar System overnight when the calculations revealed a distance of almost 2 billion miles from the Sun. Herschel became famous overnight and went on to greater fame as the originator of the general catalogs in use by deep sky observers, the discoverer of binary star systems and the discoverer of radiation beyond the visible spectrum.

As a planet, Uranus is not a good target for an observer. It just reaches naked-eye visibility at magnitude 5.7 but is barely bright enough to be noticed, a fact that caused confusion and consternation when a search of historical documents found that the first Astronomer Royal, John Flamsteed recorded it as the star 34 Tauri in 1678 in his first sweeps of the sky, and saw it six times afterward without realizing its true nature! Similarly, Tobias Mayer and the French astronomer Le Monnier also spotted Uranus but failed to note it. If such luminaries can make such a mistake, then average observers can be forgiven if they consider Uranus a less than pleasing object.

Telescopically, the object does show a small disc shining at about sixth magnitude at opposition. Under good seeing conditions it can exhibit a slight green color, which is due to a high haze of methane smog in its upper atmosphere. Of its satellites and ring system, the modestly equipped amateur will see nothing, and will have to content himself or herself with observing the slow movement of the planet against the stars over the course of the night or of several nights. So small is the disc of Uranus that it can be mistaken for a star if given a cursory glance, so once the field containing the planet is identified, sweep around it using a higher power eyepiece until the object becomes obvious.

The planet is an ice giant, although in comparison with planets such as Saturn and Jupiter the term "giant" can be rendered moot. Uranus is only five times larger than Earth, so it is hardly surprising that it is a difficult visual object given its size and extreme distance at almost 3 billion km. Any astronomy magazine or software program will enable the observer to ascertain co-ordinates to find this elusive object. We shall cover observing this planet with filters together with its neighbor Neptune.

Neptune is an elusive planet that requires a telescope to be seen properly, although binoculars will reveal the planet as a star-like object, lacking resolution. Discovered in 1846, Neptune has the great distinction of having been discovered "mathematically" before being seen visually, and the tale of its discovery is worthy of note.

Shortly after the discovery of Uranus by Herschel, observers noticed that it did not follow its predicted pattern around the Sun. Astronomers hastily calculated that Uranus was slowing down in its orbit due to the gravitational pull of an unseen body beyond it. Telescopes searched the heavens in vain, but nothing was found, and the problem was deemed irresolvable due to the intricacies of the mathematics and the lack of observational data. A young graduate at Cambridge, John Couch Adams, undertook the calculations as to the mass, orbit, size and position of this unseen planet—a task that took him 5 years.

Once the position was determined, Adams went to see the Astronomer Royal, George Bidell Airey, at Greenwich, hoping that Airey would initiate a search for the planet. However, Airey was a singularly pedantic and unimaginative individual, a good administrator, but a lackluster astronomer who dismissed Adams with excuses. Adams returned to Cambridge, where he tried in vain to interest James Challis, the director of the Cambridge Observatory, to hunt for his new planet. Unknown to Adams, Urbain Jean Joseph le Verrier in Paris had also calculated the position of the new planet, and run into similar difficulties with observers who dismissed his calculations. Stung by their obtuseness, Le Verrier sent his notes to the Berlin observatory under the directorship of Johann Enke. The night the Germans received the notes, the observatory's young observer, Heinrich D'Arrest, under the supervision of Enke and the chief observer Johann Galle, stepped up to the telescope, turned it to le Verrier's calculated position and within minutes noted the planet! Neptune had been found and another body added to the solar system.

The resultant cataclysm over the true "discoverer" embroiled scientists and astronomers alike for some time.

As with Uranus, Neptune had inadvertently been noticed before. A thorough search through historical records reveals that Galileo had recorded the object in 1612 as a bright star in the same telescopic field of view as Jupiter, and Jerome Lalande had noted the planet as a star in 1795. So attuned to not observing anything out of the ordinary were astronomers at the time that Neptune actually appears as a star on the celestial atlas of K.L. Harding in 1822!

Despite the furor of its discovery, Neptune is even more insignificant visually than Uranus. The planet is 4.5 times the size of Earth, yet is over 4 billion km from the Sun, a lonely world skirting the darkness of deep space. Visually, it shines at eighth magnitude, the only major planet of the Solar System that absolutely requires a telescope, and reveals a small disc shining with a pale blue hue. Its major moon Triton, an object with the coldest surface in the Solar System at a temperature of −235 °C, cannot be seen with small amateur telescopes, and the whole system is rather a disappointment, although finding the faint disc is an achievement in itself.

Being rather remote worlds, little was known scientifically about either of these planets until the flyby of the spacecraft *Voyager* in 1986 (Uranus) and 1989 (Neptune). The spacecraft revealed them to be astonishingly different worlds, more ice giants than gas giants. Due to their isolation, few writers have attempted to situate stories on or around such worlds, with the notable exception of Olaf Stapledon, who in 1930 penned one of the most famous SF tales of all time, *Last and First Men*. The story details the evolution and eventual destruction of humankind, but contains the interesting idea that 1 day people would learn how to modify themselves so that they could live in the inhospitable environment of the outer Solar System as the Sun expands to become a red giant star. Stapledon's "last" men, the nineteenth of their race, now live on Neptune, rising and falling in the harsh atmosphere almost like fish in the sea. It is an interesting idea in exploring extreme life environments, one which many scientists are beginning to ponder.

Until summer 2006, Pluto, one of the outer members of the Solar System, was a planet. However, a meeting of the International Astronomical Union in Prague demoted Pluto, making it part of a group of "dwarf planets," including Ceres, in the inner Asteroid Belt, Quaoar and Eris in the Kuiper Belt, a system of icy and rocky debris at the outer edge of the Solar System beyond Neptune. These Kuiper Belt objects are extremely difficult to observe, and the task is not something that many amateurs undertake. Nevertheless, some amateurs will attempt to spot Pluto during its opposition, though the use of filters of any kind will be very limiting, as Pluto generally shines at magnitude 13.8 at its brightest.

Like Neptune, Pluto was discovered after a "mathematical" search for an object that perturbed the orbits of Uranus and Neptune. Not much hope was held of ever finding a planet this remote from the Sun, and the problem remained until in 1929, Lowell Observatory in Arizona hired a talented young amateur astronomer by the name of Clyde Tombaugh to perform a photographic search. Tombaugh struggled with comparisons of photographic plates, eventually comparing the positions of over one million stars until finally, in February 1930, he found the small body that astronomers had largely forgotten. At an average distance of 8 billion km from the Sun and a diameter of 3,000 km, Pluto is tiny and insignificant visually and requires a telescope of at least 250-mm aperture to view it. Finder charts are printed in astronomical magazines and online for those interested in pursuing this distant object.

Filters for Observing Uranus and Neptune

The presence of high haze layers of methane in both atmospheres and the small angular sizes of each body make observing these planets difficult at best. Although Uranus and Neptune are still fairly bright at magnitudes 5.7 and 8.1, respectively, they subtend tiny discs even at opposition. Uranus is just 4 arc seconds in diameter, and Neptune is 2.1 arc seconds in diameter. Visually, this is about the size of the Great Red Spot and some of the oval storms on Jupiter!

As both planets are predominantly green and blue green, then the use of the yellow Wratten No. 8 or No. 15 may make some difference to your observing, or, if using a telescope of 300-mm aperture try the Wratten No. 25 deep red to see if any cloud features swim into view. However, when one considers that Uranus as observed by the *Voyager* spacecraft was unremittingly bland and featureless and Neptune revealed subtle cloud belts but is a tiny visual object then any filter use that reduces light transmission is going to affect the image more than the filter can compensate for.

Though it is not recommended here, observers with large aperture telescopes are welcome to try a high magnification combination of the following filters to attempt any resolution of features, with the proviso that tiny Neptune is the better target.

Wratten No	Color	Features
No. 15	Yellow	Possible high ice clouds on Neptune, deepening of planetary color
No. 25	Green	Possible high ice clouds on Neptune, deepening of planetary color

Comets: Visitors from Deep Space

The sight of a bright comet, its tail streaming across the heavens, is an incredible though rare sight. However, when one considers that annually there are about 40 comets visible to astronomers, by what standard do we determine how rare these events are? Simply put—any comet that is actually visible to the public or is mentioned in the national press is a rare event!

Most comets are decidedly uninteresting, even seen through a powerful telescope, and few and far between are the beautiful comets like Hale-Bopp or comet PanSTARRS 2013 to name two recent ones that delighted sky watchers.

From the point of view of the novice, comet hunting is a highly intensive and time-consuming search for the ephemeral and undistinguished. Comets vary greatly in brightness, size and visibility. One must have an intimate knowledge of the night sky and an almost photographic memory to remember the positions of thousands of faint stars in order to distinguish if any newcomer appears among them. From this point of view, comet searching is left to the dedicated amateur, or more commonly to automated telescopes involved in monitoring the activity of near-Earth objects.

Nevertheless, it can be instructive to scan the heavens from time to time just in case. Keeping a weekly eye on the comet newsletters on websites such as www.harvard.edu is a good indication of forthcoming spectacular comets that would merit observation. Coupled to this, understanding the ephemerides and locating the comet on a map and in the skies is an excellent practical demonstration of one's skills.

Comets are highly important, as they contain pristine elements and molecules from the formation of the Solar System. They are in fact time capsules, enabling astronomers to gauge what sorts of materials were in the solar nebula out of which the Sun and planets formed and how such materials have evolved in the 5 billion years since. Recent scientific research focuses on these exotic materials, as many comets not only contain water in the form of ice but also pre-biotic molecules that could have rained down on Earth and the other planets of our Solar System and contributed to the rise of life. Indeed, some have gone as far as to speculate that comets may well contain either bacteria or viruses in addition to pre-biotic matter. This is an interesting scenario that at present has little supporting evidence but nevertheless has stirred the astronomical community. If one is fortunate enough to observe a comet, perhaps it would be good to ponder the make-up of debris that it leaves behind.

This debris is frequently seen as meteor showers and some are fixed at particular times of the year such as the Geminids in December and the Perseids in August. These meteors take their name from the constellations they appear to "radiate" from, and the two mentioned here are the best of the year. Not to be forgotten are the Leonids in November and the Orionids in October, the stream associated with comet Halley. However, meteors are not a subject for filtered observing and are best seen with the naked eye.

In ancient times, comets were thought to be the harbingers of doom, ill fated, with a rather baleful aspect. The philosopher Aristotle believed comets to be foul exhalations in the atmosphere, all part of the degradation of an Earth at the rank center of the universe and removed from heavenly perfection. Subsequently, astrologers foretold the rise and fall of kingdoms by the appearance of such bodies, and many cultures occupied their astrologers and astronomers in watching the skies for the appearance of these bad-omened celestial objects in order to forewarn others. The Chinese have records going back to 200 BC, detailing observations of "hairy stars," and the most famous of all comets—Halley—features on the Bayeaux tapestry of 1066, interpreted by the court of William the Conqueror as bad luck for

Harold Godwinson the English king. Comets, by their rarity and high station, only presaged bad luck for the rulers of the day as Shakespeare acknowledged in *Julius Ceasar* (Act II, Scene II):

> When beggars die, no comets are seen,
> The heavens themselves blaze forth the death of princes

The famous astronomer Tycho Brahe began to set the scientific record straight when his observations of the bright comet of 1577 enabled him to discover that comets were outside of the atmosphere. This discovery made the Solar System a dynamic and evolving place and shattered the ideal of immutability that Aristotle demanded of the heavens. Following on several historical observations, Sir Edmund Halley performed calculations based on Newton's law of gravity that tied the observations to a single comet returning on a prescribed orbit. Comets became part of the predictable, mechanistic universe despite their rather disastrous aspect, and modern science has continued to dispel the comet myth.

Comets have a few features that are worth noting. The bright head, or "coma," of the comet is a cloud of icy and gaseous material immediately surrounding the nucleus, which is generally a very small rocky/icy body just a few km in size. The coma can be over 1 million km in diameter, and streaming behind it will be a lustrous tail (if we are fortunate) that is divided into two components, the dust tail and the ion tail. The dust tail lives up to its name as the physical silicate material streaming away from the comet in the solar wind, and the ion tail is the ionized gaseous material that reacts to changes in the solar magnetosphere and can disconnect and re-grow as the comet heads around the Sun. Figure 4.14 shows comet PanSTARRS 2012 as a typical example of such a body.

Observing comets can be done with either binoculars or a small telescope, depending on the brightness and size of the comet. Comets, due to their very nature, are unpredictable, and their quoted magnitudes on ephemerides may not be what the observer actually witnesses. This is due to their nature—comets are basically icy snowballs of varied sizes interspersed with rock and dust. The current scientific model of a "dirty snowball" was first proposed by Fred Whipple in the late 1950s and has stood the test of time and scrutiny; Halley was intensively investigated in 1985/1986, the European probe *Giotto* discovered that the nucleus was in fact a dirty, potato-shaped chunk of icy debris orbiting the Sun and spewing out gases at unpredictable rates. The melting of any cometary nucleus of indeterminate size thus releases gases and dust at an unspecified rate, relating directly to the erratic nature of foretelling cometary magnitudes. Comet ISON of 2013 is a case in point here where it was portrayed in the media as the "comet of the century," only to break up and disappear as it rounded the Sun!

These heavenly visitors are purported to come from an area on the periphery of the Solar System, about 40,000 astronomical units away from the Sun. This area, a spherical capsule around the Sun, is now known as the Oort Cloud, after the Dutch astronomer Jan Oort, who postulated the existence of a body of comets at great distance surrounding our parent star. These bodies would be formed at the

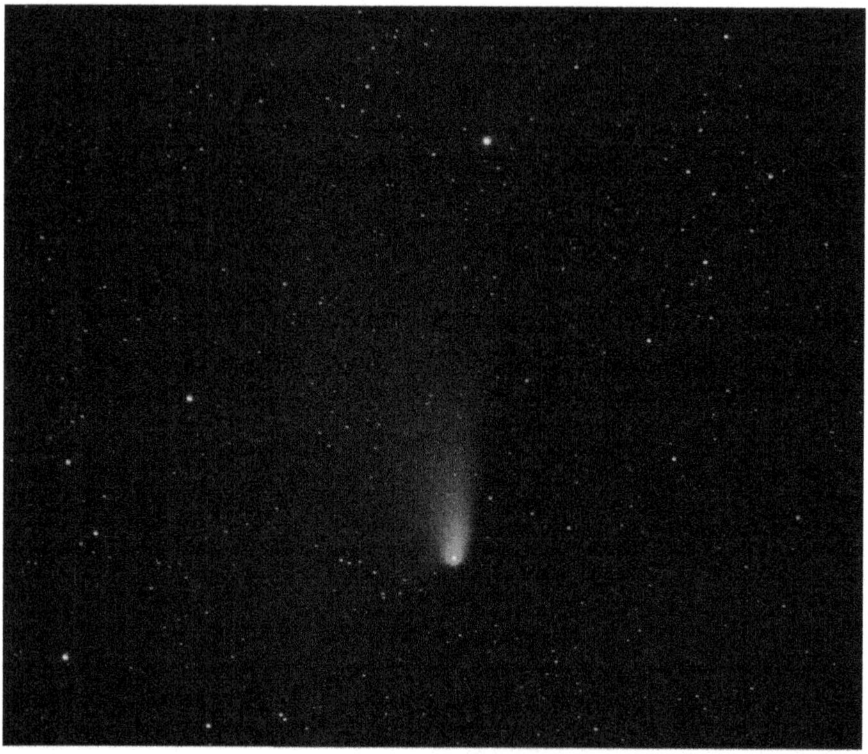

Fig. 4.14 Comet PanSTARRS

earliest time in our Solar System's history, while the solar nebula was still largely spherical and the Sun had yet to form. Although we still have no direct evidence for the Oort Cloud, the high inclination of cometary orbits, so different from the smooth plane of the planets, has convinced most astronomers of its reality. The recent discovery of a half-way house between the planets and the Oort Cloud, the Kuiper Belt, has only added to the increasing amount of evidence in favor of the Oort Cloud hypothesis.

It is thought that the short-period comets (up to 250 years) come from the Kuiper Belt, while the long period comets come from the Oort Cloud itself. Periodic comets have rather varied timescales, the shortest being that of Comet Encke at 3.3 years, Halley at 76 years and Swift-Tuttle at 120 years. Comet McNaught-Russell, featured above, will never return to the Solar System, and neither will comet NEAT, since both have hyperbolic orbits rather than elliptical ones and will escape the Sun's gravity and wander the galaxy. The fabulous comet Hale-Bopp of 1997 will return in 4,200 years' time. Clearly comet watching is a rather long and laborious process!

Filters for Comet Observation

Observe a comet carefully over as long a period as possible to see any changing features, which are generally observed visually without filter help. Nonetheless there is a "comet" filter in the market that may be a useful tool for observing.

The filter is a narrowband one rather than the colored correction or longpass filters we have so far been discussing. Comet filters have a narrow bandpass of just 25 nm that is centered around 510 nm in the green part of the spectrum and essentially isolates the oxygen component of the comet tail in addition to that of cyanogen lines in the spectrum at 511 and 514 nm. This filter, usually known as the Swann band filter, after the Scottish physicist William Swann who identified several spectral features of hydrocarbon fuels in 1856. This filter picks out the gaseous components of the tail, and any comet observer will know from experience how difficult it can be to see structure and detail in cometary tails. This filter therefore fills a vital niche in the market, and its use is recommended by such observers as Howard Brewington and David Levy.

Comet hunting requires dedication beyond the pale and lots of patience. Sweeping the horizon after sunset and working your way upward in an overlapping pattern takes some practice, but once one is confident enough, the search just may be worthwhile. If the comet filter above can add to your observing experience, then it is a valuable addition to your filter toolkit. Whenever one of these visitors from the outer Solar System deigns to visit us, it is worthwhile to capture the apparition by DSLR camera or web-cam as the above filter will also aid enhancement in imaging.

A bright comet is an extremely rewarding sight, one that is not easily forgotten by those fortunate enough to see one of these celestial visitors. The table below gives the details of a useful comet filter.

Filter	Bandpass	Features
Swann band	501–526 nm	Gaseous components of comet tails and coma

Conclusion

It is hoped that the above information gives the reader the impetus to go and use Wratten filters to explore the Solar System. These filters are generally inexpensive and are a great addition to observing through any telescope and will reveal previously unseen details if used carefully.

However, there is one Solar System object that we have not covered so far—the Sun. Observing our nearest star requires completely different filters to those described here and so it is to a description of the types and uses of solar filters that we now turn.

Further Reading

Benton J (2005) Saturn and how to observe it. Springer, New York
Grego P (2012) Mars and how to observe it. Springer, New York
McNally J (2012) Jupiter and how to observe it. Springer, New York
North G (2012) Observing the solar system. The modern observers guide. Cambridge University Press, Cambridge, UK

Chapter 5

Using Filters to Observe the Sun

The brightest object in the heavens and the most obvious, yet sometimes overlooked, celestial object is of course the Sun, our very own star. The Sun is the closest such object to us and therefore a hotbed of research whereby all other stars are measured. From an observational point of view, observing the Sun can be particularly rewarding, as the typical work that can be performed gives insight into the underlying structure and physical fundamentals of all stars.

As a body, the Sun is an incandescent mass of luminous gas 1.3 million km in diameter lying at a distance of 149 million km from Earth. The Sun consists of hydrogen (75 %), helium (24 %) and 1 % everything else on the Periodic Table, although given its surface temperature of 5,700 K, some elements are going to be hard to distinguish spectroscopically, with the result that we know only 62 of the 92 elements that are actually present in the Sun. In volume the Sun is so large that the body could hold over 300,000 planets the size of our celestial home! Such is the mass of the Sun that the SF writer Isaac Asimov once quipped that the Solar System consists of the Sun, Jupiter and debris. Overall, the Sun accounts for 98 % by mass of everything in our Solar System, a fact that at once renders us rather insignificant.

The Sun shines by converting hydrogen into helium in the core, where the temperature is close to 15 million K. In this conversion, which is known as hydrogen fusion, four hydrogen protons are used to make one helium nucleus in a process called the proton–proton chain.

When we look at the Sun, we are looking at a part of the atmosphere known as the photosphere. It is upon this surface that most of the drama available to modest 'scopes is played out. From the novice's point of view, only the photosphere concerns us, unless your 'scope is equipped with a device such as a Hα filter that will enable the observer to also see features in the chromosphere, the high temperature

M. Griffiths, *Choosing and Using Astronomical Filters*, The Patrick Moore Practical Astronomy Series, DOI 10.1007/978-1-4939-1044-1_5,
© Springer Science+Business Media New York 2015

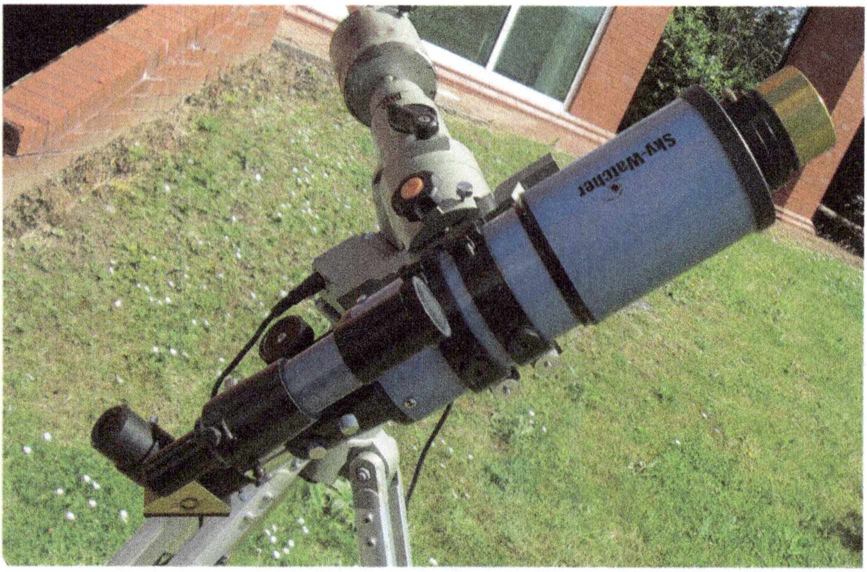

Fig. 5.1 Coronado Solarmax two-stage system

region of thin gas above the photosphere. We shall discuss the filters necessary for solar observing shortly.

Needless to say, but worth repeating as a warning is: **Never look at the sun with the naked eye or look at the sun through optical aids without the correct filters**. Ignoring this warning could result in blindness. However, with the correct filters solar observation is a very rewarding activity, and much can be gained from looking at the Sun in white light, in Calcium and Hα light. These filters reveal a remarkable amount of detail on the solar sphere that can be captured using DSLR cameras and webcams or simply just visually observing the disk. In many cases solar observers prefer to use a Hα telescope with a webcam such as a Phillips TouCam or a DMK camera. Once the video is captured it can be put through a program such as *Registax* to stack and enhance the image before moving a file into *Photoshop* for colorization if necessary.

Many solar observers have their own dedicated telescopes that are custom made by companies such as Coronado, Solar Scopes or Lunt, and these 'scopes can cost a small fortune but are well worth the investment if you wish to maximize your solar observation. Figure 5.1 shows a Coronado Solarmax Hα two-stage system. Most observers will settle for the filters at the lower end of the market that give quick accessible views of the Sun, revealing sunspots and faculae that are detailed and acceptable, especially if one is at a public event where solar observation under controlled and safe conditions can be maximized.

There is much to be seen on the solar disc, so it may be instructive to have a summary of the sorts of features the observer can expect to see on the solar surface before we move on to using filters to see them.

Solar Activity and Detail

The following is not a complete catalog of everything that can be seen on the Sun, but these features are the most obvious to the visual observer. Imaging the Sun will also show the same details but with added aspects that depend on the imaging system used and the skills one has with *Maxim DL, Photoshop* or *Registax.*

The one feature that stands out very well in white light, Hα and Ca II observing is the surface phenomenon known as *solar granulation*. These are convective cells that are best seen in apertures over 75 mm and are a lovely feature to view in the green light of a solar continuum filter. Each of these solar cells is an area of hot rising gas, typically about 1,100 km across, and have a cooler edge or "channel" of descending gas, giving them a darker edge and delineating each cell one from another.

Seen in sharp relief against the backdrop of solar granulation or just the solar disc are the *sunspots*. These are the Sun's most obvious feature, but are not always visible, as their appearance will depend on the intensity of the Sun's magnetic field and its timescale within the solar cycle, which generally ranges across 11 years from maximum to minimum. Spots come in groups or in two's (a result of their magnetic polarity—north and south), although one may see an individual spot either led or followed by a disturbed area of the photosphere. Sunspots are areas where the magnetic lines of force deep within the Sun are squeezed together until lines with similar charges meet and begin to repel one another. When this happens, the lines break through the Sun's photosphere, cooling the area and giving the photosphere a dark appearance by contrast, as their temperature is approximately 4,200 K in relation to the average photospheric temperature of 5,700 K.

Looking closely at the spots the observer will notice that there is a dark area surrounded by a ring of gaseous material that is darker than the photosphere, but not as dark as the spot itself. The spots have an *umbra*—the dark center—and a *penumbra*—the lighter surround, and are often seen in association with *faculae*—white cloud-like areas that can be seen to good effect near the edge, or limb, of the Sun. Seeing details can be difficult, as features in the umbra and penumbra can be as small as a fraction of an arc second, which equate to features of a few hundred to 1,000 km.

Even at times where the sunspot activity is at a relative minimum, there should be some spots on the visible disc. Indeed, in 2003, one of the largest and most energetic sunspot groups ever recorded was visible for several weeks in October/November, resulting in great outflows of gas and charged particles that gave great auroral displays.

Generally visible on the disc are features known as *pores*. These are tiny dark spots on the photosphere, generally smaller than 2,000 km in size, and often have fairly short lifetimes, disappearing within a few hours. Pores occasionally form where several granulation channels meet and can occasionally precede the development of sunspots.

Also associated with sunspots are areas known as *plages*. Plages are patchy, cloudlike, bright areas on the solar disk, usually found near active regions and can last for several days. As they have a strange sandy appearance under high magnification they are names after the French word for "beach." Plages are generally rather irregular in shape and variable in brightness, and they delineate areas of vertical emerging or reconnecting magnetic fields, which is why they are often associated with the magnetic regions around sunspots.

As everything on the solar surface is dominated by the interaction of the hot ionized gas with the Sun's internal magnetic field, plages are often to be found where there are *emerging flux regions*. These are areas on the Sun where a magnetic dipole, or flux tube, is surfacing on the disc and can produce a bipolar sunspot group. They are predominantly visible in Hα as a small oval area of bright plages or spicules. Occasionally the magnetic poles of these emerging flux regions are marked by pores or small developing sunspots.

Continuing our examination of areas associated with sunspots, features known as *ephemeral regions* live up to their name as limited magnetic dipoles with lifetimes of about a day that contain no sunspots. Ephemeral regions can develop anywhere on the Sun, but are more common at mid and lower solar latitudes. They appear as small brighter elements in the chromospheric network but are fainter than plages.

Very similar to plages are the *faculae*, patchy white-light bright areas very like clouds in the photosphere, but usually not visible in Hα, which are best observed when close to the limb of the solar disc, where they stand out against the limb darkening. Faculae are most often found near active regions or where one is about to form, and can last well after the sunspots in any active regions have decayed. They are best seen in blue light and so the use of a Ca II filter will bring these out to advantage.

Most of the above features can be seen with white light filters, though obviously the degree of detail that can be glimpsed will depend on factors of atmospheric seeing, quality of telescope and filtration and the resolution of the image. The following features are generally seen in Hα, or in the case of prominences, visually only during a solar eclipse.

The most astounding features on the limb of the Sun during Hα observing are without doubt the *prominences*. Prominences are huge clouds of gas streaming along magnetic field lines, and they create a complex of clouds and streamers of gas that ascend into the chromosphere. They generally come in two broad classes: active and quiescent, which we shall look at in detail a little later. Associated with them and seen as a dark feathery line against the bright solar disc are the *filaments*, which often mark areas of magnetic shearing and are very similar to prominences. Both features twist and turn continuously, and it is very difficult to see a prominence or filament in the same place on 2 consecutive days, as they change their size and appearance continuously, usually lasting several hours.

These prominences and filaments are also part of the *chromospheric network*, formed by magnetic field lines related to supergranulation cells in the photosphere below, which is a network of long sinuous chains of tiny low-contrast and thin looking brighter areas on the disc that also have the name of *filigree*. These filigree features are also found in plages—if one can maintain a high magnification and extend over much of the solar disc in a Hα telescope. These bright points often have darker *spicules*, areas of long thin streamers of gas that rise out of them and make the actual network harder to see.

One of the most transient features of the solar disc and one that many observers long to see is the *solar flare*. These are emission features that can last from just a few minutes to over 4 h. Solar flares are a rapid and violent release of energy in the chromosphere due to extreme magnetic field stress and can occasionally result in material leaving the Sun in the form of a *coronal mass ejection*. When such ejections occur, and the released material is heading directly toward Earth, the aurora, which usually is only seen at high latitudes, can become a relatively low latitude event. There are many websites that give warning of such events, and even the British Meteorological Office is now hosting regular space weather updates.

Associated with solar flares are events known as *surges*. A surge is simply a transient prominence produced by flares or very active regions, appearing as a jet of gas rising up from the surface of the Sun. Such ejected gas will often fall back onto the Sun tending to follow magnetic field lines, while at other times it will rise and disperse and therefore fade from view.

With so much detail visible on the Sun's disc it is unsurprising that observers tend to invest in solar telescopes with either Hα or Ca II capability. Additional visual treats in Hα or Ca II will be dealt with under the relevant observing section. However, as these are expensive items and not within the economic range of all, let us first examine solar filters and their observational uses, moving from low-cost, low-detail materials to more professional filters.

Mylar and Baader Paper Filters

One of the best filters for any solar observing is either Mylar or Baader paper. Mylar is an aluminized plastic that stops 99.9 % of the Sun's light penetrating the telescope, rendering a nice clean white light image with a slight blue tinge as can be seen from Fig. 5.2. Baader paper, created for solar viewing by the Baader Planetarium in Germany, or Baader Astro Solar safety film (to give it its correct name) is almost similar in appearance and is a thin sheet of foil that can be cut to size to fit the apertures of a telescope or binoculars and fixed in place. Both filters allow light through with a peak wavelength of 540 nm, right in the center of the Sun's optical output. It must be stressed that despite its application to binoculars and spotting 'scopes, a telescope is best used for solar observing, as the image size is larger and reveals more detail.

What the observer will see with such filters is a clean, round disk of light—the photosphere of the Sun, shown here in Fig. 5.2a, upon which should be displayed such details as sunspots and faculae. The sunspots may have enough detail to reveal

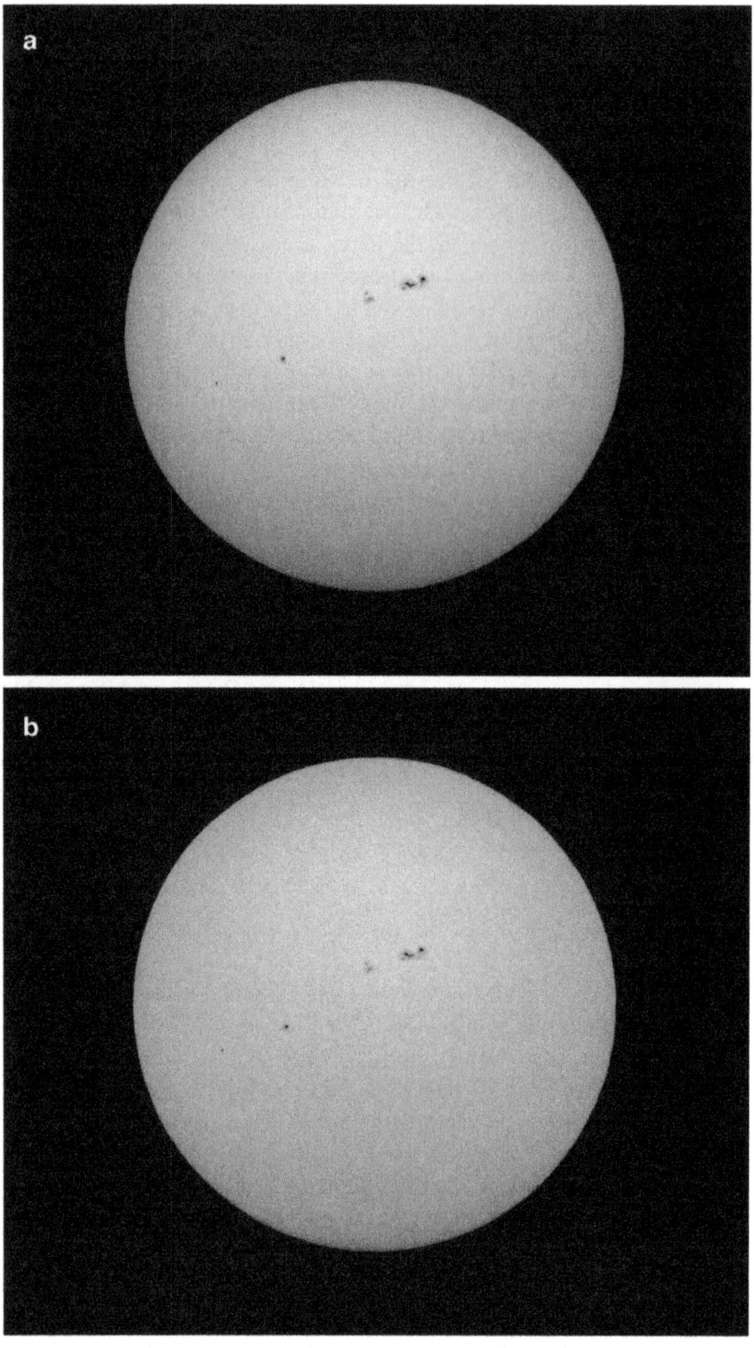

Fig. 5.2 Baader paper (**a**) and Mylar Sun (**b**)

the inner dark penumbra and the lighter outer penumbra and areas of activity such as the whitish faculae that appear in areas of intense solar magnetic activity.

One does not need a very large telescope to observe these details. Most observers use a 60-mm telescope or smaller, as the Sun is very large and relatively close as an astronomical object. Coronado, Solar Scopes and Lunt telescopes have small ranges of between 40 and 100 mm, which are more than adequate to observe our nearest star.

Additionally, it is very easy to take photographs through Mylar or Baader paper and then reduce them in *Photoshop* to gain better contrast and even add color to them. This can be done with a monochrome shot through such filters by the simple expedient of going into Photoshop and going to **image > mode > RGB color**. Once RGB color is ticked then go to **enhance > adjust color > adjust hue/saturation** and tick the **colorize** box at bottom right. Use the sliders then to get the correct color hue and lightness and saturation before going off to adjust the image under **enhance** in other ways. Other software packages will have slightly different menus, but the method is similar. Figure 5.3a show the difference such coloration can make.

Herschel Wedge

The Herschel wedge was invented in 1830 by John Herschel, son of the Royal Astronomical Society President, William Herschel. The wedge itself is a prism that allows the passage of light through a neutral density filter, and the heat and excess light from the Sun is dispersed out of the back of the housing, leaving about 4 % or less to enter the eyepiece. This was certainly a great improvement over his father's method of solar observing, which involved jugs of claret being placed between eyepiece and flat!

There is an immediate caveat with the Herschel wedge, however. It is designed to be used with a refractor only. Reflecting telescopes collect and focus too much IR light through the eyepiece and can even crack the prism in the wedge and prove dangerous to the sight of the observer. The wedge also requires the additional use of neutral density filters for observing and photography, though these can be purchased anyway. The wedge decreases the amount of light entering the system but does not narrow the spectral sequence down to a few bandwidths, so the spectral profile of the Sun remains unchanged in contrast to other 'white light' filters that do block out certain spectral wavelengths.

Additional filters can of course be used with the wedge, and many observers use calcium K line and solar continuum filters (see below) with this tool to access more detail on the solar surface, though most enjoy the view and take photographs simply by adding a neutral density filter, as can be seen from Fig. 5.4, an image of a partial solar eclipse taken by the author using a Herschel wedge.

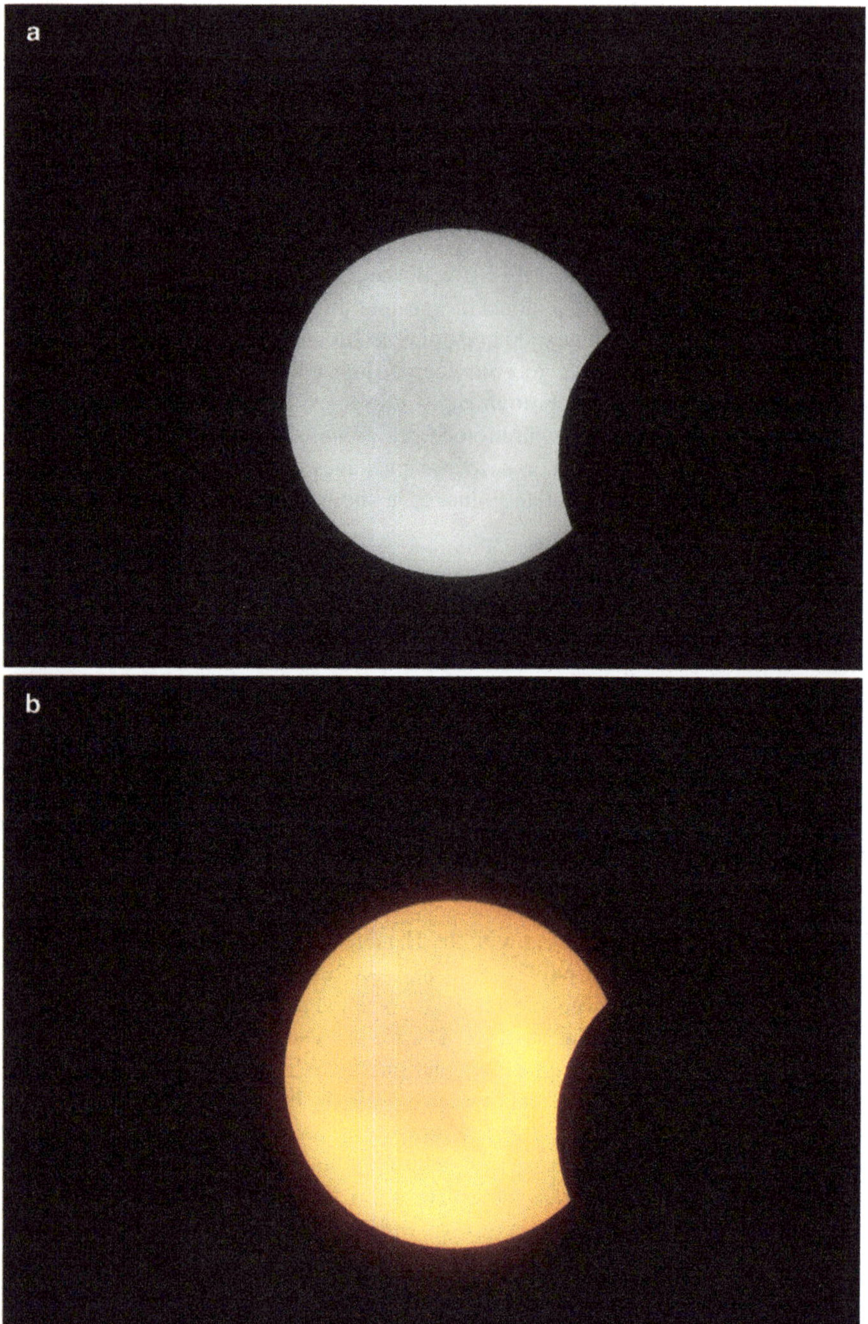

Fig. 5.3 Baader paper eclipse (**a**) and *Photoshop* colored version (**b**) (image taken by Martin Griffiths)

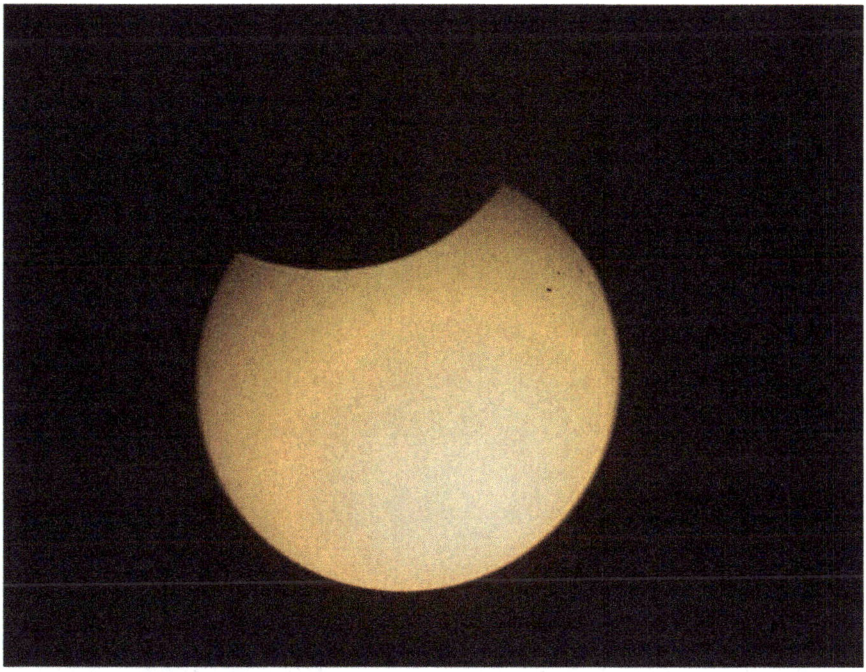

Fig. 5.4 Partial eclipse with Herschel wedge

Solar Continuum Filters

The solar continuum filter is designed to use in conjunction with the Baader Astro Solar filter or the Herschel wedge, and additional surface details as contrast and light transfer are enhanced. The filter has a bandpass of 530–545 nm in the optical spectrum, where details of faculae, penumbra, granulation and a host of finer detail can be seen.

Observers often remark on the greenish tinge to the view in the eyepiece—it is very strange to see a green Sun, as can be imagined from looking at Fig. 5.5! However, the bandpass is close to the sensitivity of the human eye, and the effect of contrast and detail plus the seeming suppression of atmospheric turbulence make this filter choice a must for serious solar observation. With the continuum filter the granulation of the solar surface is easy to see, and the areas around sunspots take on an almost 3-D appearance, with the spots themselves demonstrating some depth and clarity between the umbra and penumbra. In higher magnifications on a small telescope (depending on the optical quality of the telescope) it is possible to see the striated pattern of the gases in the penumbra twisting in the magnetic field.

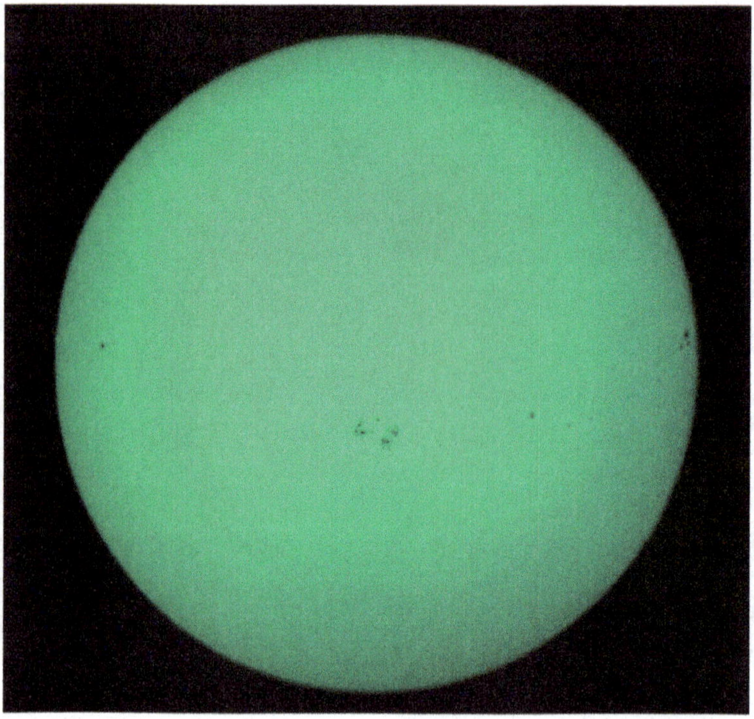

Fig. 5.5 Sun in green light

Calcium Filters

These filters render a very blue image when used visually as the peak wavelengths in calcium are just at the visible/UV edge of the spectrum. Such filters are suitable for photographic work and should always be used in conjunction with a Baader Astro Solar filter. The Calcium H and K line filters allow the observer to see the photosphere in detail at wavelengths of 396 and 393 nm, respectively, revealing minutiae in the Sun's upper photosphere and chromosphere, where flares and plages can be seen to advantage. These filters are made with reflectors in mind as the blue light is collected well by such instruments. These filters are also known as Lyot monochromators, wherein they use the phenomena of birefringence to cancel out particular wavelengths of light, leaving the wavelength that the observer wishes to use. Such filters are specialist items and can be expensive, but are well worth the investment.

Coronado has built a reasonably priced calcium solar telescope in the personal solar telescope (PST) arena, but the observer needs to wear some sort of hood to

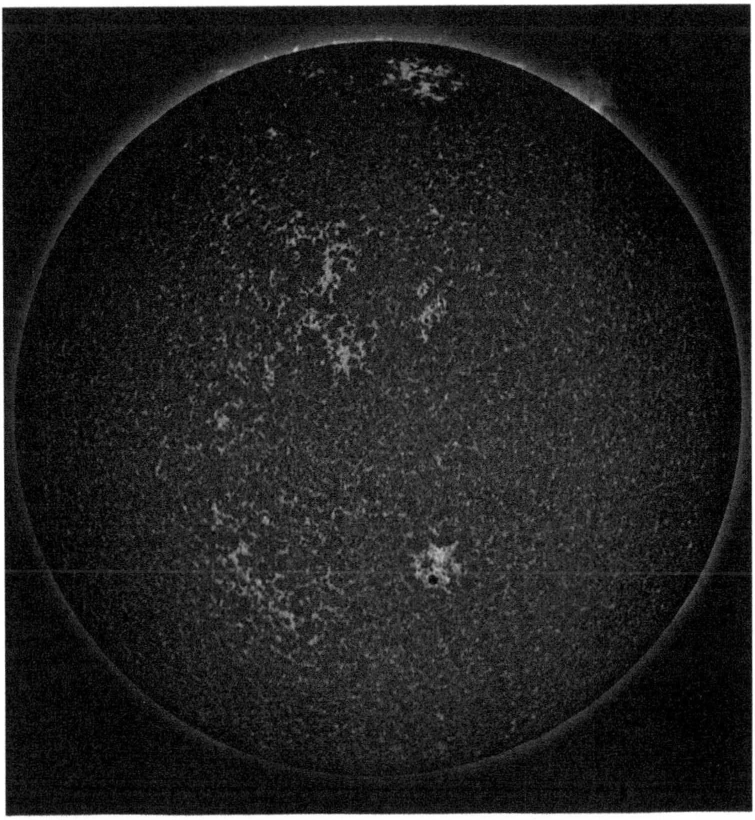

Fig. 5.6 The Sun in calcium K (*Credit*: NASA)

allow a deep shadow to cover the 'scope and head, as the images tend to be dim when viewed directly. Other manufacturers also make calcium filters or specially designed telescopes used for imaging the Sun in this light, but it becomes an expensive hobby at that point!

Nevertheless, once the eye has adjusted to the blue and relatively dim image a variety of fine detail can be seen with such an instrument. Spots and flares can be seen in detail, as can the white plages in the lower chromosphere, the solar granulation, filaments and even prominences on the solar limb. Figure 5.6 shows a typical shot using such a filter. With a calcium filter and Astro-solar filter on a larger telescope much finer detail can be seen and recorded on the disk as the fine granulation comes into view. In the K line the sensitivity to the solar magnetic field is prominent, with weak fields showing up as dark areas and stronger field lines rendering the area whiter before very strong fields render the image dark again, especially in the areas of sunspot umbra.

Hα Filters

This filter is the filter of choice for most dedicated solar observers. The hydrogen alpha (Hα) system reveals the Sun in wavelengths of red light centered at 656 nm and produces a dazzling display of orange-red light visually. This is the Sun's most active spectral line, and so observers equipped with Hα telescopes gain immediate access to sunspots, granulation features, filaments, plages, faculae and prominences on the solar limb—a wealth of features in the photosphere and chromosphere of the Sun.

Some of these filters come as a two-stage system such as the Coronado Solarmax, which can then be attached to an ordinary refractor for visual and photographic work or come as a single integrated system that you just couple a webcam to such as the Solarscope's SV systems. There are many excellent Hα systems on the market, and even the Coronado PST shows a lovely accessible image that its larger brothers will greatly add to in detail.

The image shown in Fig. 5.7 is a typical representation of the solar surface and chromosphere as seen with a Hα telescope. This photograph is a reduction of 1,500 webcam images using a DMK 21 monochrome camera and processed in *Registax* before moving the picture over to *Photoshop* for coloration. The details visible in

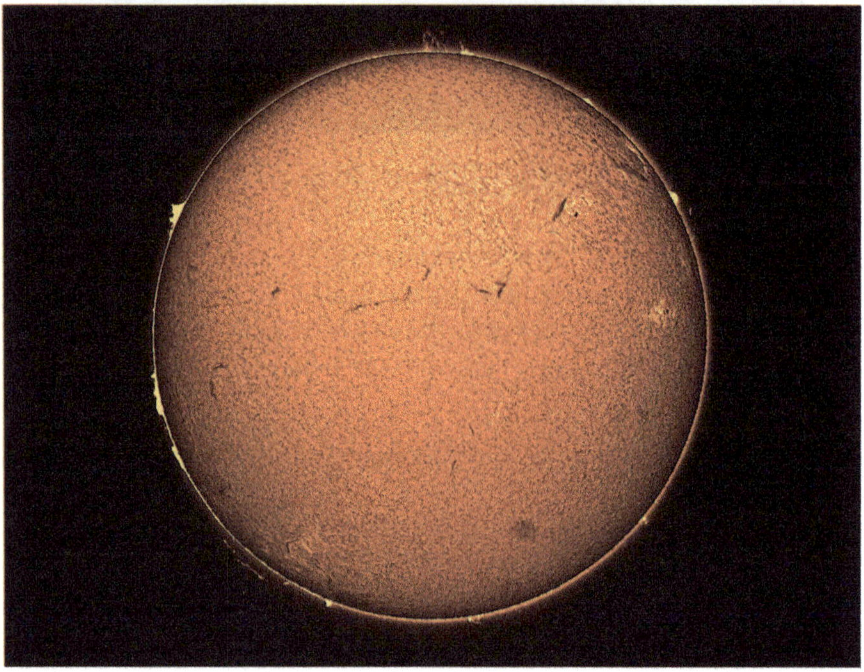

Fig. 5.7 The Sun in Hα (Photograph by Andy Burns of Wiltshire Astronomical Society)

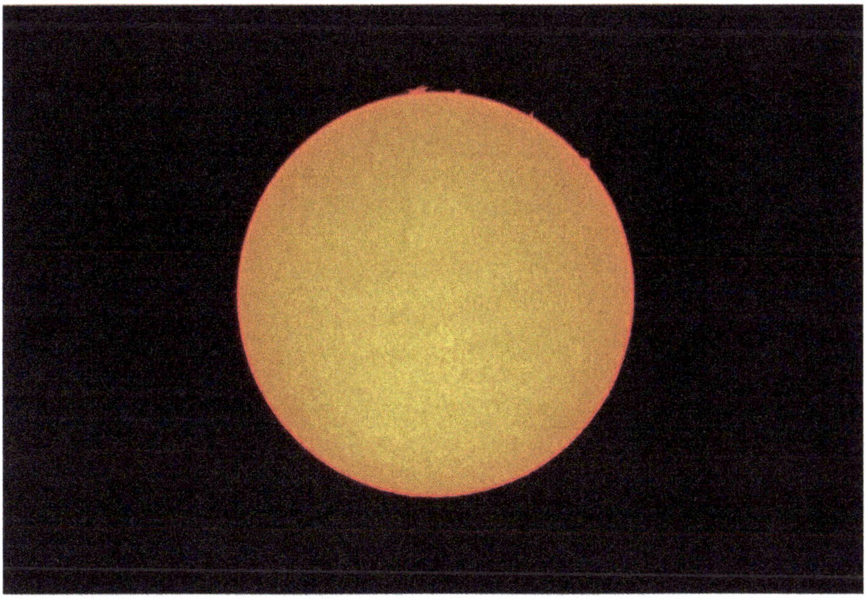

Fig. 5.8 The Sun and prominences

Hα are plainly visible in this wonderful photograph taken by Andy Burns of the Wiltshire Astronomical Society.

In a telescope with high resolution other features appear, such as spicules and fibrils, areas of the Sun that are small and filamentary in nature in addition to ethereal regions such as Ellerman bombs, which are very transient small areas in flux regions where the magnetic field of the Sun breaks through its surface. Chromospheric networks, pores and flux regions are all visible in detail as long as the telescope is either dedicated to Hα observations as in the case of Coronado, Solarscope and Lunt telescopes or is equipped with the Coronado Solarmax filter or equivalent.

Bracketing exposures on a DSLR camera can capture many details, and then combining the frames in *Photoshop* as can be seen from Fig. 5.8 here. Prominences and the granulated surface of the Sun with areas of faculae are clearly visible, even though such shots do not provide the level of detail that can be captured with a DMK camera or equivalent.

Observing the Sun with White Light Filters

Though some of the features visible on the solar surface have been covered briefly above, it would be in order to spend some time discussing in detail the activity visible on the surface of the Sun.

Fig. 5.9 Sunspot umbra and penumbra

The white light filters we have already discussed, Mylar, Baader paper and the use of a Herschel wedge, will show the most obvious solar features, such as sunspots, limb darkening and faculae. Depending on the quality of the telescope, it is also possible to see granulation features in white light, too. With such filters the observer can follow the progress of sunspots across the disk and make records by means of photography or sketches of the positions of the spots on the disk. The observer will note the relative rapidity of any changes to the spots' shape, size and their number as they progress across the solar photosphere, giving a clue to the differential solar rotation. It was using this form of observation, noting the drift of sunspots, that led Galileo to conclude in 1610 that the Sun was rotating on its axis.

Observing the Sun over a few hours will demonstrate the dynamics of spot groups, as they almost seem to change before your eyes. Attempting to discern detail in the spots is quite a challenge with white light, as the spots may be small and in white light do not reveal the complexities of the hot gas as well as a Ca II filter or the Hα filter does.

A sharp eyed observer will note that the solar surface is broken down into small areas surrounded by a darker halo, almost as if one was observing the pitted surface of an orange or a bowl of rice. This is solar granulation, the result of convection currents bringing hot gases of the interior up to the photosphere. These granulation cells can be thousands of km across and represent another white light feature that can be recorded. Figure 5.9 shows the umbra and penumbra of a sunspot group in white light along with some granulation.

Observing sunspot groups will enable you to gauge the rotation of the Sun, which is an interesting exercise demonstrating the gaseous composition of our local star. This can be achieved by merely counting the days when a large sunspot group appears on the disk and drifts to the edge. Observing the groups from edge to edge and making a rough calculation (×2) can then give the solar rotation in days.

With such filters the observer can also attempt to gauge the size of each spot. This measurement can be performed using any filter in practice, though it is the clear cut appearance of spots in a Mylar or Baader film filter that makes the process a little easier for the observer.

The diameter of the Sun is 30 arc minutes, so one can then make either a guess or an accurate measurement with a micrometer of the diameter of a single spot or even a group of spots. This can be estimated in arc minutes and then converted to arc seconds. Once this is done, one can use the small angle formula to work out the size of the spot. The formula is:

$$\mathrm{D} = a\mathrm{d} / Rad$$

where D is the real diameter of the spot in km, *a* is the spot's apparent size in arc seconds (″), d is the Earth–Sun distance (149.6 million km) and *Rad* is the amount of arc seconds in a radian (57.3° or 206,265″).

If the observer measures a spot to be 85 arc seconds in diameter the sum will then be:

$$85 \times 149{,}600{,}000\,\mathrm{km} = 1.27 \times 10^{10}\,\mathrm{km}$$

So,

$$1.27 \times 10^{10}\,\mathrm{km} \div 206{,}265 = 61{,}571\,\mathrm{km}$$

Earth's diameter is 12,715 km, so dividing the above result by the diameter of Earth one gets the figure of 4.8—or 5 times the size of Earth to significant figures! A spot that takes up approximately 1 arc minute on the solar disk, and therefore looks quite small, will actually be something very significant indeed!

Using the results of this simple formula at public events always tends to make people gasp in amazement; the general public are impressed when comparisons such as this can be made, as it brings home the vast size of astronomical objects in a context that people can grasp. Couple this to the fact that the Sun can be seen safely using modest equipment, and pavement astronomy comes to life—and is not confined to the cold and dark of the evening.

One of the things to look for while observing sunspots is the so-called Wilson Effect, named after Dr. A. Wilson, a Glasgow astronomer of the eighteenth century. This is an apparent narrowing of the penumbra when the spot is on the limb of the Sun. Strangely, one would expect the umbral edge nearest the limb to be drawn out and narrowed like the images of lunar craters on the Moon when they are at unfavorable librations; instead the opposite is true. The umbral edge nearest the center of the Sun is narrowed, and the one nearest the edge remains the same as if it were at the center of the disc. Look for this in your observations.

If you wish to include some additional information in your log, then attempt to classify the sunspots according to the Macintosh system, which is rather complex and has 60 different classifications. However, for simplicity try arranging your sunspot recordings according to the following table, which is based on the Macintosh data.

1. Small spot or group with no penumbra
2. Small group of spots with no penumbra but displaying symmetrical pattern
3. Bipolar group with largest members displaying penumbra
4. Bipolar group in which major spots have penumbra
5. Very large bipolar group with major spots displaying complex penumbra and smaller spots displaying penumbra
6. Largest bipolar groups with complex penumbra and small spots
7. Largest bipolar groups with complex penumbra and no small spots
8. Large spot with complex penumbra surrounded by random small spots
9. Single spot with penumbra

As the reader can discern, even with Mylar or Baader paper filters there is much to see on the solar surface. Narrowband filters with a specific bandpass will show a host of details that are unobservable to those equipped with white light filters, and it is to these that we now turn.

Observing the Sun with Narrowband Filters

One of the most productive filters for solar observing is the Calcium II filter. This narrowband filter typically has a 8-nm bandpass centered around 395 nm in the blue part of the solar spectrum and gives sharp details in the Fraunhofer K line. The filters can be quite expensive, and some are dual-stage filters, which also include the white light filters such as Baader paper or Mylar. The Ca II filter shows features on the solar photosphere and in the hot overlying region, the chromosphere.

The calcium lines in the Sun can be used to predict where extensive sunspot formation may take place, as the faculae noticeable at these wavelengths generally point to future sunspot activity on the disk. Through the Ca II filter, moderately strong magnetic activity shows up as bright areas on the Sun's disc, but the strongest magnetically active areas are the spots themselves, which always appear dark. Faculae and plages are a few hundred Kelvin hotter than the photosphere and stand out appreciably with these filters. The calcium filters also reveal fantastic details of the supergranulation on the solar surface and fine bright spots on a quiet Sun known as 'K grains,' but good resolution and patience are needed to observe these features, as they are very transient, lasting about 5 min in all, as the convection cells on the solar surface rise and deposit their energies.

The filter can also be used to look for Ellerman bombs, named after Ferdinand Ellerman, who studied these phenomena in the early twentieth century and concluded that they were microflares in the chromosphere associated with magnetic

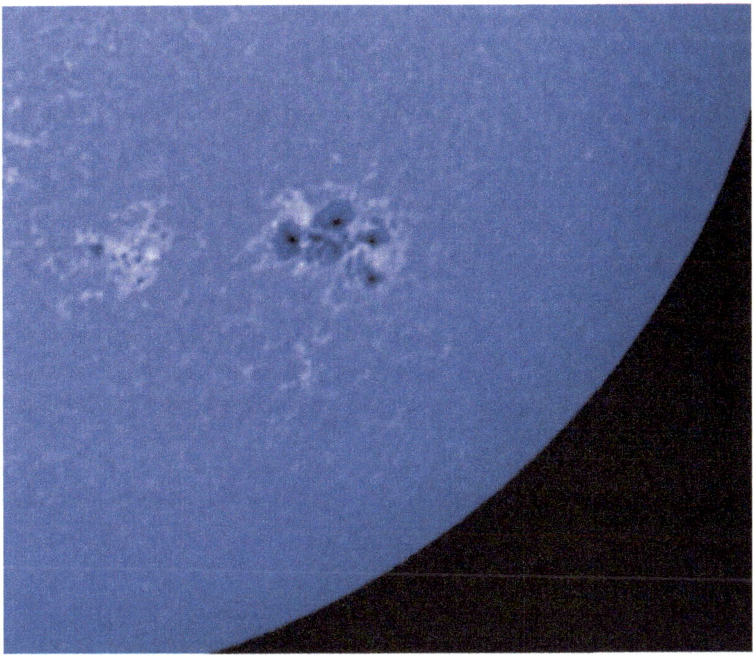

Fig. 5.10 Spots and plages in Ca II (*Credit*: NASA)

field reconnections, where two opposing streams of ionized material collide with a brief flare of light and energy. Obviously one has to be patient and alert to see these relatively rare occurrences. Observers using this filter can make out details that are visible in Hα but at a different, more energetic end of the spectrum.

The Ca II filter gives a very dark impression of the solar surface, but once in use the detail and clarity of such areas as granulation, plages, sunspots and areas with high magnetic activity are almost second to none and can be captured using a DSLR or DMK camera quite easily as can be seen from the details of plages, spots and granulation in Fig. 5.10.

Ca II vs. Hα

The Ca II filter is a great addition to the solar observer's toolkit. Most observers however still prefer the old standby of the Hα filter for its versatility and the amazing number of features it reveals in the photosphere and lower chromosphere. With a narrow bandpass of 12 nm centered around the Balmer transition at 656 nm, this wonderful filter reveals a very active solar surface overlain by features such as prominences, filaments, sunspots, spicules and plages and the grainy solar granulation in unprecedented detail and at a wavelength that is bright and comfortable with a lovely orange hue that is pleasing to the eye. There are various types of Hα filters that can be directly fitted to the camera, fitted to the eyepiece or come as specialist telescope attachments, as in the Coronado two-stage system we saw above and in the Lunt, Solarscope and Coronado solar telescopes which are obtainable through most astronomical suppliers.

Using an Hα filter enables the observer to see features in great detail, especially in types of prominences around the solar limb and the dark feathery filaments that cross the solar disc. The pattern of granulation visible in a Hα filter is amazingly detailed and really does look like the surface of an orange to the visual observer. There is a very 3D effect when looking at the sun in Hα that one does not get with any other filter, as one really can pick out the photospheric and chromospheric effects easily and can photograph these details with some degree of comfort.

Details in the prominences around the limb and their differing types can be a wonder to behold and baffling to catalog or describe. However, solar prominences can be classed into two major groups. This visual cataloging was first attempted by the solar astronomer Harold Zirin, perhaps better known as "Captain Corona" to his students, who thought so much of him and his work that they thought he was some kind of superhero! The Zirin groups are:

Zirin Class I (Quiescent)

A: Hedgerow
B: Curtain, flame or fan
C: Arch and platform arch
D: Cap and irregular arch
E: Eruption

Zirin Class II (Active)

F: Eruptive prominence
G: Surge
H: Spray
I: Flare loop

The majority of these designations can be seen in the accompanying Hα photograph of the Sun in Fig. 5.11. See if you can spot the types for yourself.

This gives the observer a ready guide to classifying prominences for their logs. Watching the solar prominences even over a short period of time gives a very dynamic impression and reveals the speed of the gas moving in the magnetic fields, as within a few minutes some prominences change their appearance. Using the small angle formula given earlier can also allow the observer to measure the height of the prominence above the solar limb and the extent of these moving gaseous curtains.

In addition to the above prominences, the observer may notice a patchy network of long thin chains of tiny low contrast brighter points called filigree, which are also found in plages. This is part of the chromospheric network, and it extends over much of the solar disk in Hα.

Spots can be seen in unprecedented detail in a good quality telescope with a high magnification, as the area of the penumbra breaks into streaks of gas that are

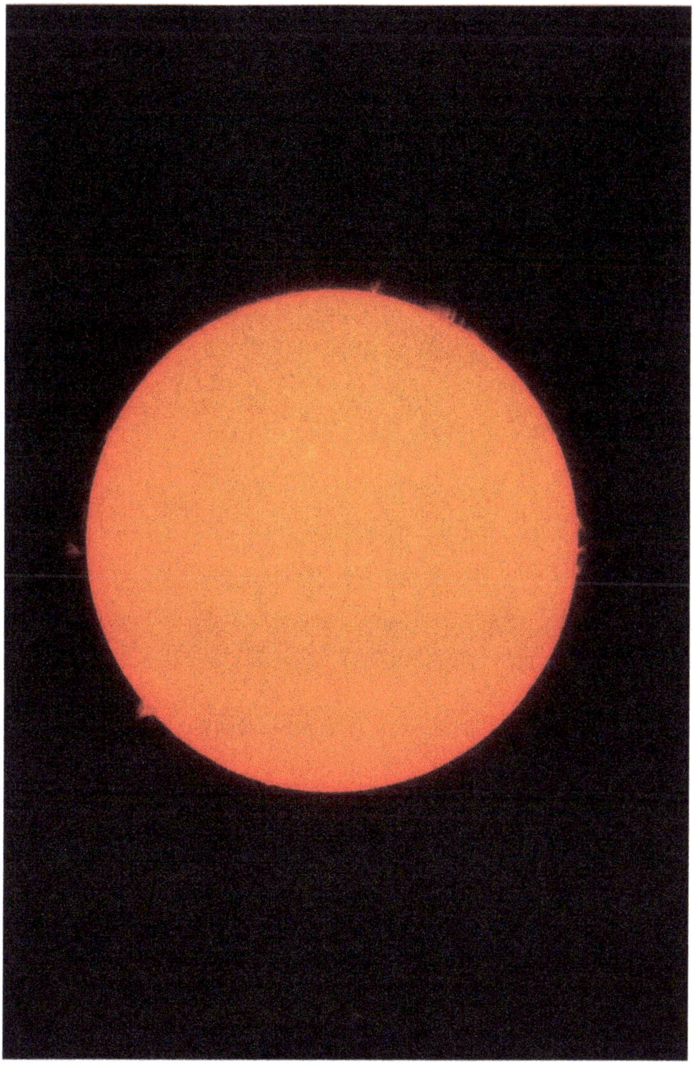

Fig. 5.11 Prominences for Zirin comparisons

almost visibly writhing in the magnetic field. Under good conditions and high magnification, details of the spicules on the solar surface can be seen, and imaging brings out remarkable detail, as can be seen from this DMK image of a sunspot and spicule in Fig. 5.12.

The use of narrowband filters transforms solar observing. Although white light filters are very useful and versatile, for sheer detail and imagery impact the use of Hα and Ca II cannot be overstated. At many public observing events the ability to show an audience a color image rather than a monochromatic one makes a world

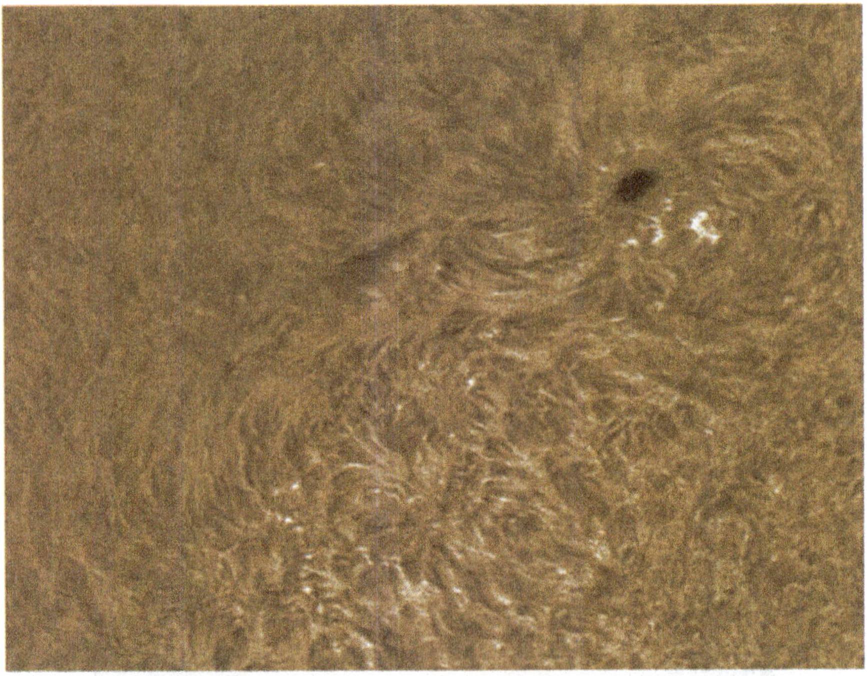

Fig. 5.12 Sunspot and spicules (photograph by Andy Burns of Wiltshire Astronomical Society)

of difference and gives the public imagination a boost, as many are unaware that one can even look at the Sun through a telescope.

For those who want to make this type of observing the focus of their activities see Springer's *Observing the Sun—a Pocket Field Guide* (2013) by Jamey Jenkins or other books of similar ilk. As a quick guide here is a table of the features that can be seen on the solar surface with the correct filters.

Filter	Bandpass (nm)	Features
Mylar	400–750	Solar granulation, sunspots, umbra and penumbra of sunspots
Baader Astro Filter	400–750	Solar granulation, sunspots, umbra and penumbra of sunspots
Herschel wedge	400–750	Solar granulation, sunspots, umbra and penumbra of sunspots
Solar continuum	530–545	Solar granulation, sunspots, umbra and penumbra of sunspots
Hα	650–660	Solar granulation, sunspots, umbra and penumbra of sunspots, prominences, faculae, filaments, chromospheric network, spicules, solar flares.
Ca II	389–397	Solar granulation, sunspots, umbra and penumbra of sunspots, prominences, faculae, filaments, chromospheric network, spicules, solar flares, plages, Ellerman bombs, K grains.

Conclusion

So far we have been covering the types of filters that can be used for visual observing. Although some of the filters previously discussed have crossover into photography, this type of activity is a more rigorous and detailed application and requires the correct types of cameras and imaging equipment as well as the ability to reduce the image by combining the results of photography through various filters into one whole image. Therefore the filtration systems used on CCD and DSLR cameras will be the subject of our next chapter.

Further Reading

Green S, Jones M (2009) Introduction to the Sun and Stars. Open University Press Milton Keynes, UK
Jenkins J (2009) The Sun and how to observe it. Springer, New York
Kitchin C (2001) Solar observing techniques. Springer, New York
Mobberly M (1999) Astronomical equipment for amateurs. Springer, New York
Pugh P (2007) Observing the Sun with Coronado telescopes. Springer, New York
Zirin H (1997) Caltech oral histories. California

Chapter 6

Filters for Astrophotography

A step up from the colored glass filters of the Wratten system are those that are especially built to allow specific wavelengths of light through that correspond to particular energies in the elements present in astronomical bodies. There are many filter manufacturers on the market, and such filters are not always a cheap item, but if you are an avid photographic astronomer, they will make a world of difference to your pictures and are worth considering.

There are filters built for the telescope, filters built for CCD imaging and even those such as made by Hutech or Astronomik that fit the front of DSLR camera lenses and come in a variety of sizes to fit the lens optic. Most of these filters are comparable in price. It obviously depends on the kind of photography or visual work the astronomer performs as to which filter is best suited to his or her needs.

Considering that most CCD cameras on the market are monochromatic, any photographer will require filters to turn their images into dazzling color and reflect the nuances of astronomical objects in a vivid and visually correct way. In other words, filters are a must to achieve this goal.

Photometric Systems

The most obvious filters that will be required are the red, green and blue ones that make color rendition in cameras possible. These three colors, two of which are primary, represent the two ends and the middle of the electromagnetic spectrum of visible light. Green (or the v-band) is used as the "middle" color, as the eye has a visual peak at 500 nm in the green part of the spectrum and represents a more accurate determination of true color in the spectrum. These filters, which have the collective name of *photometric filters*, are generally regarded as RGB or UBV and

M. Griffiths, *Choosing and Using Astronomical Filters*, The Patrick Moore Practical Astronomy Series, DOI 10.1007/978-1-4939-1044-1_6,
© Springer Science+Business Media New York 2015

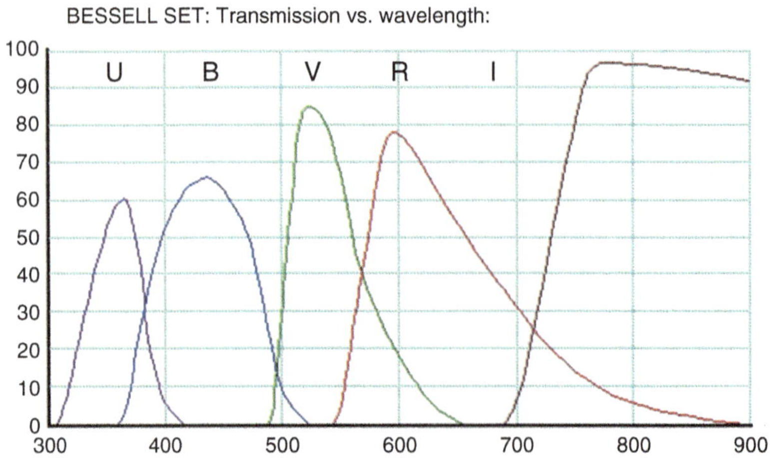

Fig. 6.1 UBVRI bandpass

are usually joined in deep sky photography by an infrared blocking filter that is either known as I or as an L band filter, thus rendering basic CCD photography of the sky an LRGB exercise or a UBVRI. We shall discuss these filters in a moment.

Filter sets were standardized by the work of Harold Johnson in the 1950s and then again by A. Cousins in the 1970s and are known collectively as the Johnson-Cousins UBVRI system. A 1990 paper by Michael Bessell in the *Journal of the Astronomical Society of the Pacific* reduced the filters of the Johnson-Cousins system into a more affordable set for budding astrophotographers by using Schott optical glass filters that, when placed together, are in everyday astronomical use and represent a versatile system that honors his name.

Filters were later added to these systems that incorporate the infrared wavelengths and are known as JHK filters, after their bandpass in the I–R part of the spectrum, where J has a bandwidth of 1.1–1.4 μm. The H filter corresponds to the H infrared line at 1.5–1.8 μm, and the K band corresponds to the infrared band at 2.0–2.4 μm. Needless to say these filters are specialist items and will not be found among the general armory available to amateur astronomers. Additionally, filter sets have been produced for specific uses at professional observatories and based upon photoelectric detection such as the Gunn-Griz system. The specific bandwidths of each UBVRI filter, based on Bessell imaging for astrophotography with DSLR and CCD cameras, can be seen here in Fig. 6.1.

There are many photometric systems in use, which are generalized into *closed systems* or *open systems.* The closed photometric systems are used by a small band of observers with a specific astronomical target or intent; these observers use a variety of filters for different purposes, and the chosen filter allows them to control the instruments and data better and maximize the consistency of their results. The Sloan Digital Sky Survey and the HIPPARCOS satellite are examples of closed photometric systems, and there are many others. In contrast, an open photometric system uses standard stars to gauge effective color and balance and are general tools that utilize standard filtration.

The wavelength that determines the bandpass of each filter type can be seen along the x-axis of the illustration. It is important to know that the SI unit of wavelength is the meter, and that 1 nm approximates to 1×10^{-9} m. An angstrom is 1×10^{-10} m and, although in use in astronomy, the nanometer is the preferred unit. Each filter will therefore cover a specific bandwidth of wavelengths in color, while the more narrowband filters such as OIII and Hα will allow small bandwidths centered on their particular wavelength (501 and 656 nm, respectively). It is the colored filters that render the image its correct appearance, and other filters tend to add to the specifics of the image or used solely to obtain detail about the object at a specified wavelength.

The common filters for UBVRI photography systems can be found in the following table. Although there are over 200 photometric systems in use, the table illustrates the most common filter components and their bandpass and corresponds to the Johnson-Cousins and the Bessell systems. Most amateur astronomers will utilize imaging systems that make full use of UBVRI plus the narrow bandwidths of the more specialized filters we shall discuss in due course.

The Johnson-Cousins photometric system

Filter letter	Bandpass (nm)	Full-width half-maximum (nm)
U	365	66
B	445	94
V	551	88
G	550	90
R	658	138
I	806	149
Z	900	125
Y	1,020	120
J	1,220	213
H	1,630	307
K	2,190	390
L	3,450	472
M	4,750	460

Using filters with a monochrome CCD camera to obtain a true color image follows on from the kind of photography that we discussed in the historical section of Chap. 1. Modern photography with DSLR cameras incorporate colored filters on the chip, known as the Bayer system, to render true color. The filter pattern on the chip is generally 50 % green, 25 % red and 25 % blue, thus gaining the most bandpass from the visible, but the filtration systems that can be added to the lenses will involve either neutral density filters or the more specialized Hα. Beyond this DSLR cameras become limited for astronomical work, and the CCD is to be preferred.

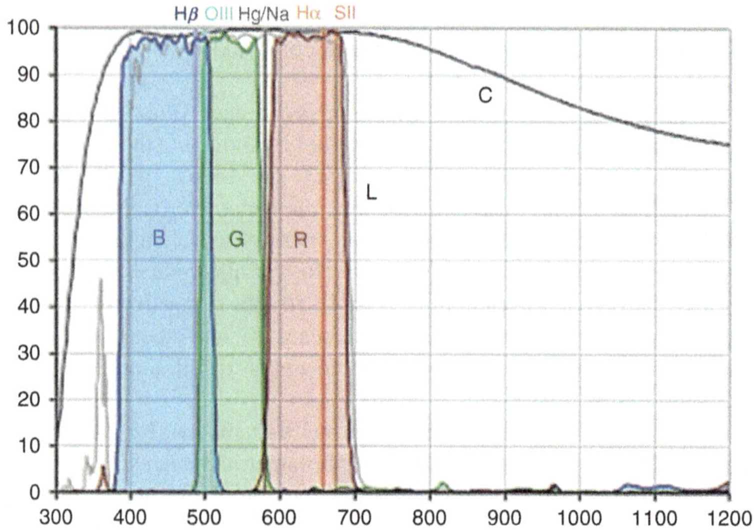

Fig. 6.2 UVB filter bandpass

For CCD imaging the standard filter sets for UBV imaging will be red, green and blue filters with the appropriate bandpass. This is generally standardized to the following bandwidths: blue (400–480 nm), V or green (480–580 nm) and red (590–700 nm), with the infrared blocking I or L band filters covering 700–900 nm, as can be seen in Fig. 6.2.

CCD chips are generally very sensitive to red and infrared, and modern filters are constructed of special Schott glass to accommodate this, which does differ slightly from the bandwidths of the Bessell system noted above, but not so much that this system is completely changed. As an example, the I band filter in the Bessell system is now generally referred to as the Ic and has a steep cut off at 900 nm, with no light transmission beyond this. It is also no longer made of Schott glass and has an additional dielectric coating to block near infrared. This is valuable for CCD chips that are sensitive to these long wavelengths, especially if studies of very red stars are a part of the observing program. However, for most photographers the minutiae of details will not be that important. All the user wants is for the filter to work at the appropriate wavelength.

Added to the UBVRI filters above will be more specific ones with particular bandpasses that enable the CCD camera to become a very useful photographic and photometric tool. Specialist filters for Hα, Hβ and OIII, in addition to light pollution reduction and ultra-high contrast filters, are widely available and are sold by most good astronomical suppliers. These filters work by allowing only a narrow selection of wavelength through and can be geared to make the most of particular objects. We shall deal with these in some detail in another section.

Useful Filters for DSLR Imaging

Although most of the above detail deals with the application to CCD imaging, similar filters can be obtained for use with the lenses or bodies of digital SLR cameras. As mentioned briefly above, manufacturers such as Hutech, Lumicon and Astronomik construct filters with both broad and narrowband ranges specifically for wide-field photography. Such filters are either a direct fit on the front of the lenses of such DSLR cameras as Canon, Nikon and Olympus, or have an adapter system to fit the lenses. With their wide field, they work very effectively at reducing light pollution and gaining resolution and image depth from suburban sites. The following photograph of the Andromeda Galaxy (M31) (Fig. 6.3) demonstrates the difference using such filters can make to the image from suburban sites. This image was taken from the light-polluted South Wales valley with a light pollution filter fitted to a 300-mm lens on a Canon 30D DSLR at f5.6 and ISO 1,600 with an exposure of 180 s and processed in *Photoshop*.

Many astronomical suppliers offer filter models that clip onto the interior of the camera just in front of the reflex mirror, or fit on the rear of the lens. These smaller filters are usually cheaper than the ones that fit the lens front if expense is a consideration, although some Canon EF-S lenses will not work with these filters, as they will not allow the lens to be connected due to its rear cell depth into the body. If you have such a camera, then such clip-in filters are best used with the filter in front of the reflex mirror and with the DSLR connected to a telescope rather than a lens.

In the same manner as the filters we shall describe below, these large filters not only reduce light pollution from sodium or mercury lamps but also reduce the atmospheric airglow from ionized oxygen at 538 and 630 nm. Most filters of this type reduce the bandpass to a range of between 440 and 530 nm.

Additionally there are the specialist filters that have application to both the DSLR/CCD camera systems, which are described in detail a little further on in this chapter. These filters, when used on DSLR cameras, generally narrow the wavelength bandpass and enable subtle features to be photographed in astronomical objects such as NGC 7293 (the large planetary nebula known as the Helix Nebula) and HII regions such as the Lagoon Nebula (Messier 8) and the North American Nebula (NGC 7000).

All filters in these specialist ranges will ensure that exposures during DSLR photography are a little longer, usually up to twice the exposure time that would be recommended without a filter. As an additional caveat, these interference filters do not give good results if the photographer is using a wide-angle lens to capture the sky. They are designed to work only on limited areas of the sky and preferentially fit telephoto lenses or in the case of the clip-in filters, with a telescope.

Obviously with such filter systems on a DSLR, the work of the photographer is limited to the dark sky. Hα, Ca II and other filters for solar imaging will not work on the DSLR, as the light transmission is far too great. Therefore, the specialist filters we examined in Chap. 5 are more useful for solar work, with a DSLR coupled via a T-mount to a filtered telescope.

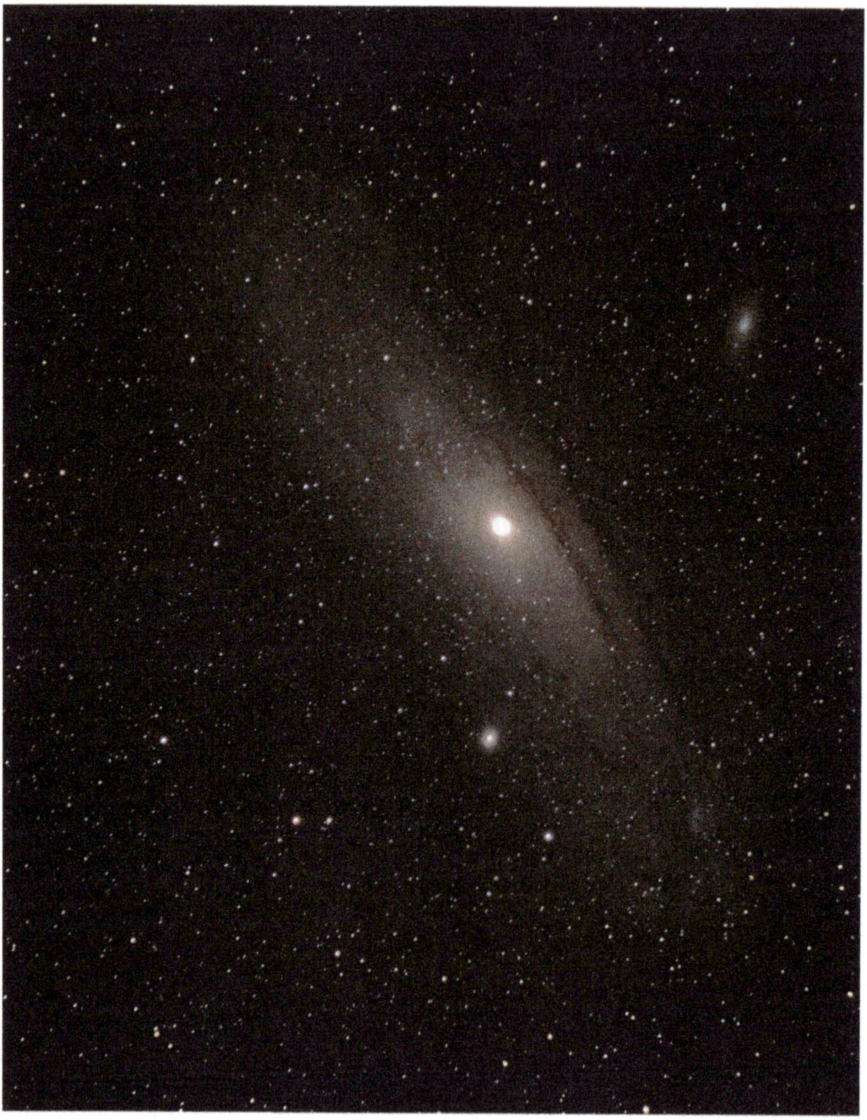

Fig. 6.3 The Andromeda Galaxy taken using an LPR filter

As an alternative to the clip or lens filters, some companies offer a "full spectrum modification" to the DSLR camera so that it can be used for both terrestrial and night studies. This usually involves modifying the chip and providing a clear glass full spectrum filter that allows both UV and IR light through and renders a high quality image that can then be used on its own to image the sky or in conjunction with other filters to gain the maximum input from astronomical sources. DSLR cameras are fitted with filters by the manufacturer that cut out extraneous UV or IR light and render colors in the visible mid-range of the spectrum. Such manufactured filters also remove good red response and, as many astronomical sources are red and have large Hα output, sensitivity to this wavelength is at a premium. However, such modified DSLR cameras are best used on reflector telescopes, as the parabolic mirror will focus all wavelengths to the same point, whereas a refractor telescope does not do this for the longer IR wavelengths. It is up to photographers to balance their needs and application to night sky photography when considering such systems.

Due to these modifications the DSLR cameras that are now available are becoming a serious alternative to the CCD camera, in the opinions of some astronomers. For the most part, though, CCD imaging is here to stay, and although filter sets for DSLR cameras have the same general features as those mentioned below, it is to the CCD systems and their appropriate filters that we now turn.

Useful Broadband Filters for CCD Imaging

Filters are an essential component for CCD imaging, though prices and quality will vary depending on location and local manufacturers. As a general rule, they can be broken down into broadband, narrowband and line filters, though these general descriptions work for DSLR cameras also.

As one might expect, it is difficult to make a recommendation with so many commercially available filters by Lumicon, Celestron, Orion and Meade. The best thing is to talk to people with experience using particular filters from these manufacturers for their viewpoints.

One final thought. It is a common misperception that filters are unnecessary if you have truly dark skies. This does a disservice to you as an astronomer and the potential of your images and the filter manufacturers, too.

In this section, we will discuss some of the commonly available broadband filters on the market today that are useful for CCD imaging before moving on to the more specialized filters that are narrowband. Broadband filters work by effectively limiting the bandpass to a specific range of wavelengths and so cut out large swathes of the EM spectrum that would otherwise be hindered by sodium or mercury vapor street lighting. Due to this they are commonly known as light pollution reduction filters (LPR). They also have the generic term nebula filters (NB), as they dim street lighting and stars while preferentially allowing the main wavelengths of Hα and OIII through—which of course are the peak wavelengths of nebulae.

In the observational list in Chap. 8, we will place LPR filters under the general rubric of NB for clarity. In Fig. 6.4 we can see a few of these broadband filters for reducing light pollution.

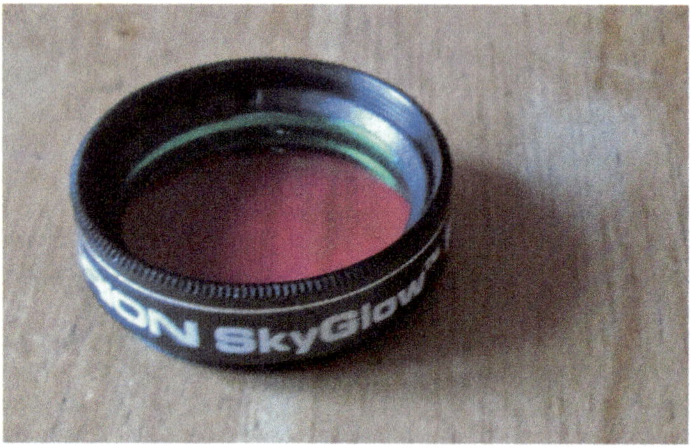

Fig. 6.4 LPR filter

UV/IR Cutoff Filters

This filter is suitable for both visual and photographic imaging and allows the wavelengths of between 490 and 680 nm through to the eye or CCD camera and allows the observer to see the object at a typical color temperature close to the wavelengths preferred by the human eye. This filter can eliminate the color fringes around bright stars and offers a 995 nm light transmission rate for most objects. It can also be used on a variety of apertures.

SkyGlow Filters

This filter is made by Orion optics and is designed to darken the background sky by blocking light transmission from streetlights that use mercury vapor. The filter enhances the wavelengths of hydrogen beta (480 nm), doubly ionized oxygen (OIII at 502 nm) and hydrogen alpha (656 nm) regions of the spectrum. These are quite versatile filters for visual or photographic work on nebula and planetary nebula, as these are common narrow wavelengths emitted by such astronomical objects.

Celestron LPR Filter

As in the case of the Skyglow filter from Orion, this filter also enhances the transmission of light in the 480, 502 and 656 nm wavelengths and is a good filter to add to any collection, especially if you are an urban astronomer or leading public observing sessions from suburban areas, though obviously the Orion or Meade filters do the same job.

Meade Broadband Filters

This filter is Meade's own LPR filter, and once again, the spectrum of wavelengths passed by this filter is identical to that of the Orion SkyGlow and the Celestron LPR filter. All this filter does is give the observer a choice as to which manufacturer they prefer to obtain their materials from, as the Celestron, Orion and Meade filters all perform a similar function.

The effect of these LPR filters can be seen here in Fig. 6.5a, b by comparing a light polluted image of M42 from an urban location with one taken at the same location with an LPR filter. Although the exposure was a little longer with the LPR filter, the clarity of the image and the darkness of the background compensate for the orange glow.

Infrared Blocking Filter

Most CCD cameras and DSLR cameras have inbuilt IR blocking filters that stop wavelengths longer than 700 nm and decrease noise in the systems of such cameras. These filters are for imaging only and are not used for visual astronomy. Check your camera specifications or contact the manufacturer if you are unsure that it is fitted with this filter.

Minus Violet Filters

At the opposite end of the spectrum in blue light you can use minus violet filters for color correction on your DSLR or CCD camera. The filter blocks all wavelengths short of 400 nm and can also be used as a visual corrector to remove flare from images of bright stars. These filters are a common component in UBVRL imaging.

Enhancing Filter

Occasionally one may come across an enhancing filter for a DSLR, which is a broadband filter that acts as an intensifier for some specific wavelengths, generally the red end of the spectrum. They are usually constructed from rare Earth elements

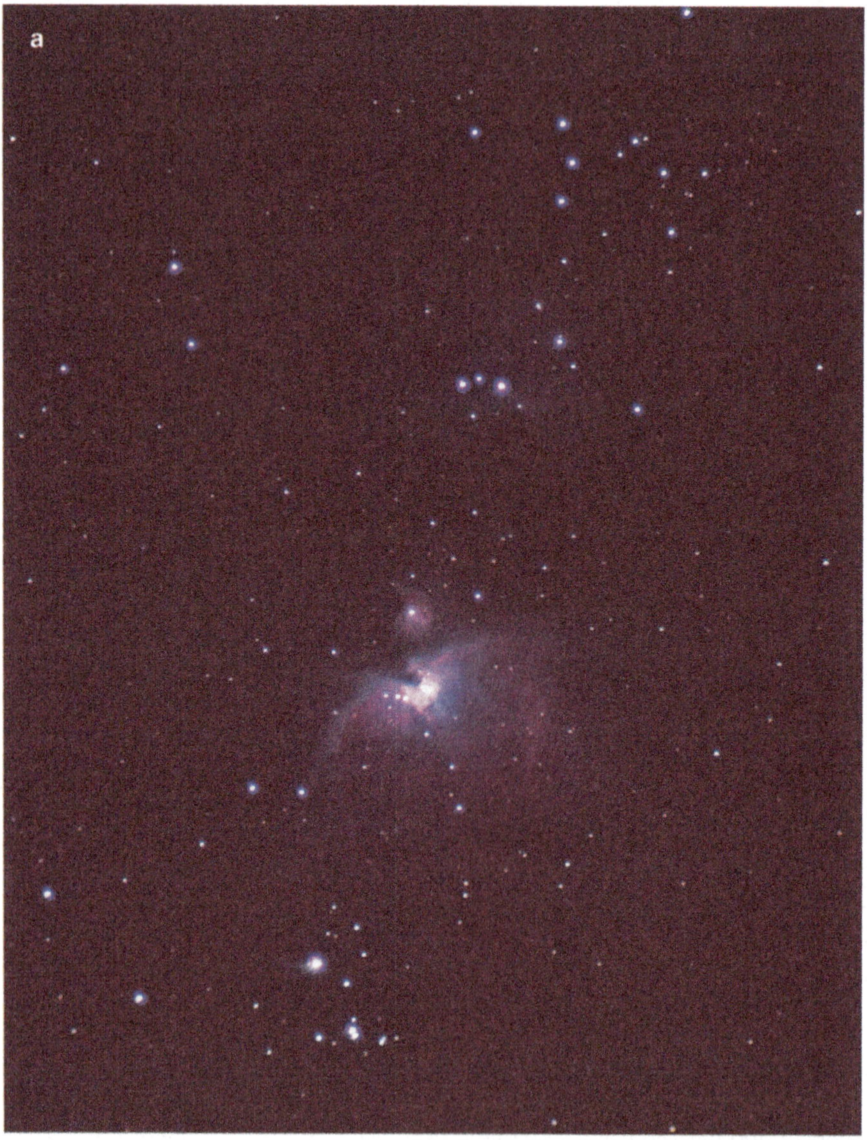

Fig. 6.5 (a) M42 and light pollution and (b) M42

such as didymium, which removes some light in the 600 nm part of the spectrum and increases the saturation of red objects but has almost no effect on other wavelengths of the spectrum. Several manufacturers, such as the well-known Hoya, sell this filter at a fraction of the cost of the narrowband filters generally preferred by photographers of the sky. This author has never used such filters and has no specific recommendation on their use.

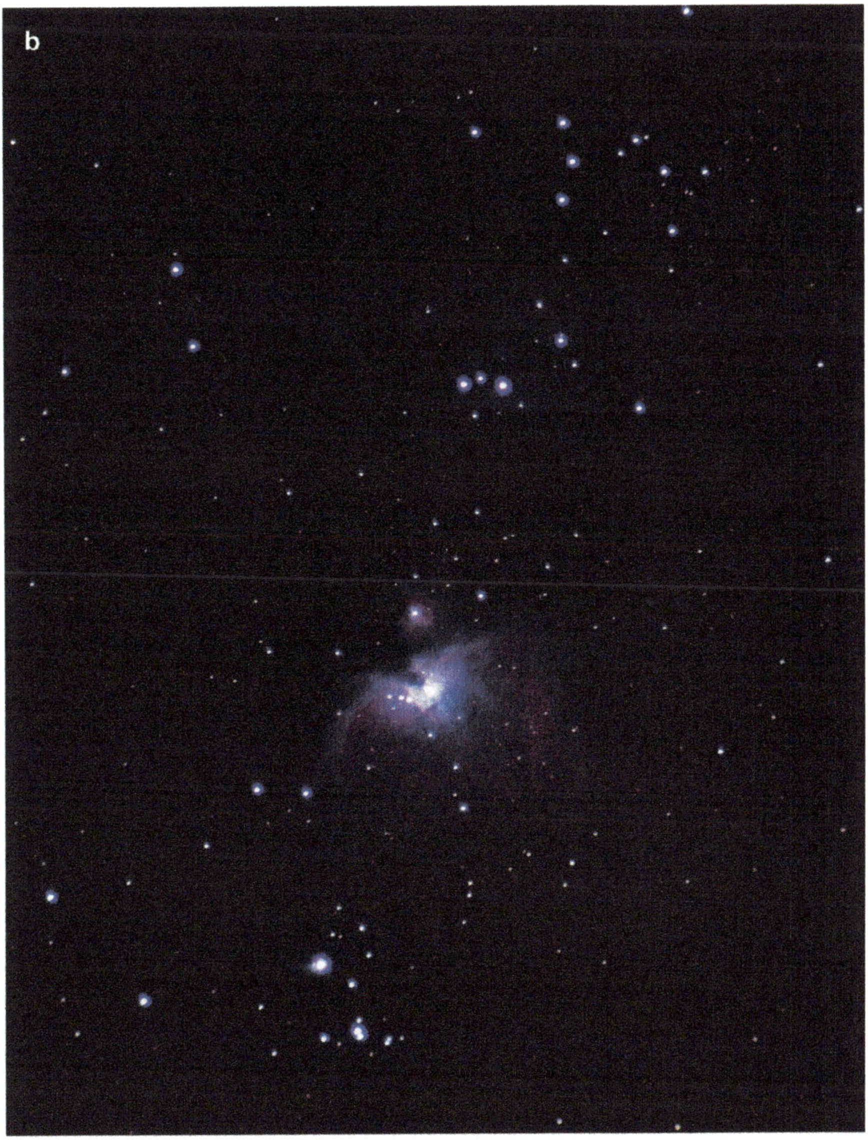

Fig. 6.5 (continued)

Narrowband Filters

There is a very wide range of manufactured narrowband filters on the market, and each type has its uses and applications. These groups of filters work to enhance contrast by letting through a narrow range of wavelengths based on the peak emission lines of particular astronomical objects.

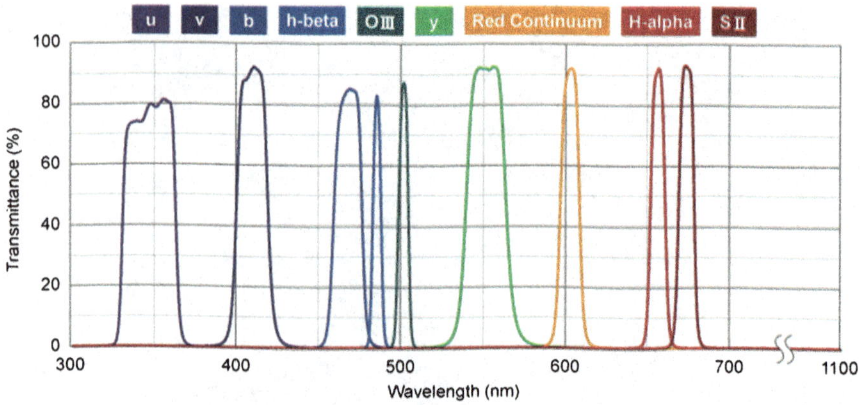

Fig. 6.6 Narrowband light transmission

One of the main advantages of these filters is that they can be used to image deep sky objects even in bright moonlight. An even bigger advantages, however, is that they can produce wonderful results even from light-polluted locations and are therefore in great demand among today's urban dwelling astrophotographers, who can produce observatory quality results without having to move their equipment to dark sky locations. Narrowband filters have revolutionized night sky photography via DSLR and CCD cameras, as fixed urban observatories now can achieve high quality imaging at relatively small expense (Fig. 6.6).

With such filters the narrow range of wavelengths that can be obtained, with decreased background noise, is known as the full-width at half-maximum (FWHM) intensity. FWHM is used to describe a measurement of the width of an object in an image, when that object does not have sharp edges—a typical astronomical source, in fact.

Narrowband filters are more difficult to manufacture, and are usually a little more expensive than their UHC or LPR cousins. Nevertheless they provide excellent contrast on most astronomical objects and provide a very good way of illustrating the correct color of such objects, an important consideration during image processing (Fig. 6.7).

Lumicon Ultra High Contrast (UHC)

This filter is narrowly focused around wavelengths that correspond to Hα at 486 nm and OIII at 502 nm, revealing nebulosity across a range of objects from emission to planetary nebulae and blocking sky glow. It works in a similar fashion to the LPR filters we saw above, but has a much more restricted range of wavelengths.

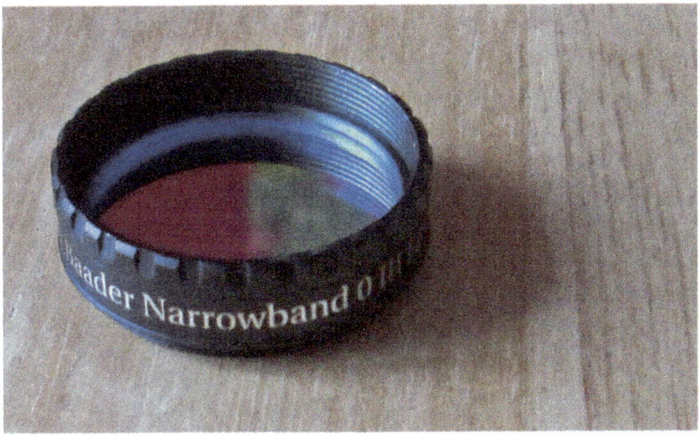

Fig. 6.7 OIII filter

Orion UltraBlock

This filter also transmits light in Hα and OIII wavelengths and blocks the typical light output from sodium or mercury vapor streetlights very much like the filter above it is a good tool for use on emission and planetary nebula.

Meade Narrowband

This filter is a little more versatile than the two mentioned above, as the Meade filter also transmits light from Hα while maintaining a block on light pollution from streetlights.

Comet Filter

This is a relative newcomer to the market, but this filter concentrates on a narrow range of wavelengths between 501 and 511 nm in the emission of carbon molecules (C_2) in the spectrum. It lets the observer record the faint details in the ionization tail of comets and provides a better contrast against which to differentiate ion tails from gas tails in these occasional visitors.

The Hydrogen (Hα) Filter

Probably the most popular narrowband filter in use, the Hα filter improves contrast between the object and background and can illuminate the delicate features of most astronomical objects a little more effectively. With a wavelength at 656 nm and bandpass of 12 nm, it is a very versatile filter and can be used on extended nebula such as HII regions very well. Hα filters are now a standard in deep sky photography, as this wavelength in the red end of the visible Balmer spectrum tends to be a preferred one among a huge range of astronomical subjects.

The Hydrogen (Hβ) Filter

This is another specialized line filter. It has a 9 nm bandpass centered around the wavelength of 486 nm in the Balmer series of the hydrogen atom transitions in visible light and is good at illuminating some low ionization features in diffuse nebulae. Many photographers use it on delicate and faint IC objects, where its range brings out the contrast between the gas and any dark dust lanes or globules. For this reason the Hα filter is commonly known as the "horsehead filter," after the wonderful dust cloud B33 in Orion.

The Oxygen (OIII) Filter

A most useful filter for planetary nebula observation is the OIII filter. This narrowband filter allows an 11-nm band of light centered on a wavelength of 500 nm through, which corresponds to the chief nebular lines in planetary nebula and are of great assistance in recording such objects. The OIII filter suppresses the light from stars within the field while prioritizing the emission lines from such nebulae, and it appreciably improves the contrast between object and background, making it spring into view visually and enhancing it greatly for astrophotography. The OIII filter is a very useful tool in the photographer's armory.

The Sulfur (SII) Filter

A filter that is coming more into use with each passing year is the SII filter. Revealing faint detail in the light of singly ionized sulfur at a wavelength of 672 nm with a FWHM bandpass of 7–10 nm depending on the manufacturer, this filter enables the photographer to capture faint detail in many nebulae and reveals the structure of feathery details in many objects at a wavelength that remained almost unexplored by amateurs until the advent of CCD cameras and filter systems.

To give the reader some flavor of the effect of each filter we include here some images taken in Hα, Hβ and OIII of the planetary nebula NGC 6302, the Bug Nebula. Planetary nebulae are wonderful objects on which to use this range of filters as they are strong emitters in Hα and OIII. The filters not only pick out relevant details in the object but faithfully render the correct color once stacked alongside RGB (BVR) filters. These images also demonstrate *why* such filters are part of the photographer's arsenal. Note that the Hα line is not well represented in planetary nebula, so when imaging it is wise to plan first which set of filters are appropriate (Fig. 6.8).

The amount of detail present here is quite remarkable across two of the images, and the combination with RGB color then renders the object as defined and close to life as possible, as can be seen from Fig. 6.9.

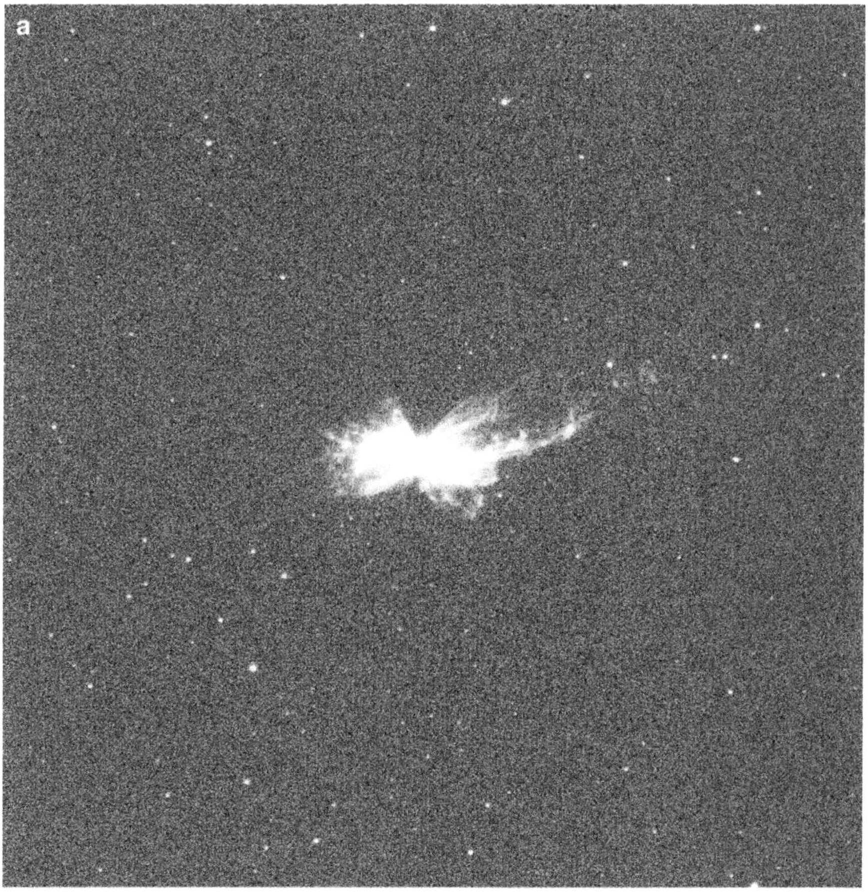

Fig. 6.8 Hα (**a**), Hβ (**b**) and OIII (**c**) images

Fig. 6.8 (continued)

When one considers that dark sky sites are an increasing rarity in most countries, and there are few astronomers within easy reach of International Dark Sky Association designated parks or reserve, a narrowband filter, by its very nature, enhances the visibility of objects with emissions at specific wavelengths and is therefore even more effective among city lights, dust and smog. Such filters also offer additional detail under moonlit conditions and so extend the amount of observing time throughout the waxing or waning of the Moon.

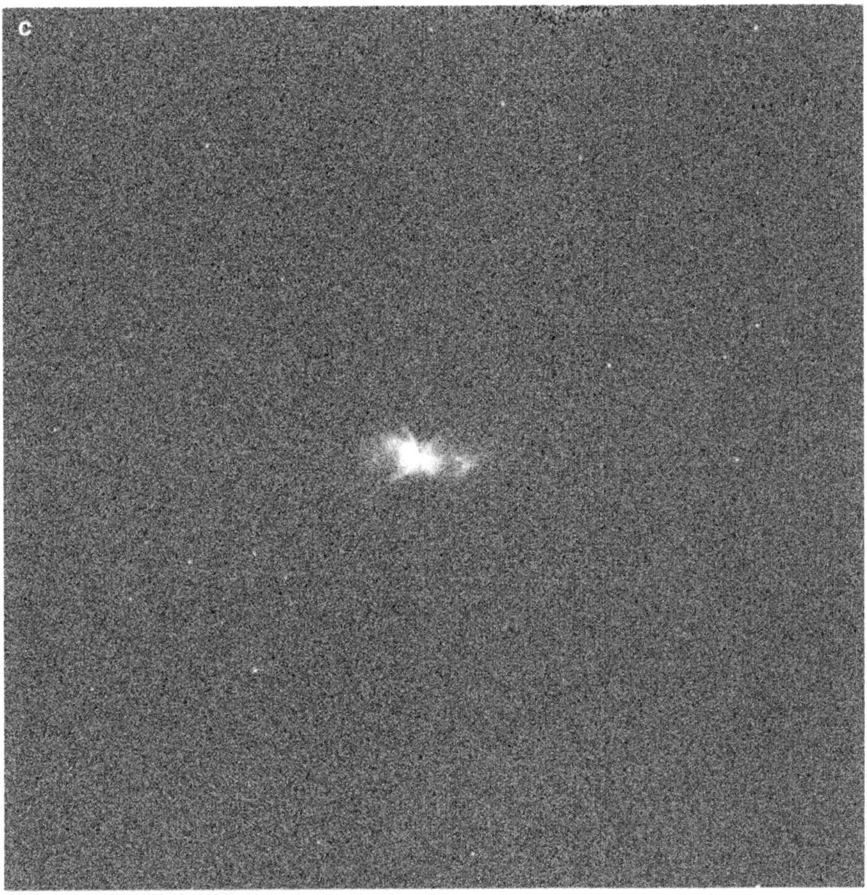

Fig. 6.8 (continued)

Filter Wheels

One excellent addition to the astrophotography toolkit is the filter wheel, which as the name suggests is a wheel-like device that holds several different filters that fits between the CCD camera and the telescope optics. It can be completely manual, requiring the user to turn the wheel onto the next filter, or it can be completely automated and run off either the telescope mount, a laptop computer or an independent power source. There are many filter wheels on the market, and they bear the usual names that are well-known and trusted, such as Atik, Andover and Starlight Express. They are generally light in weight, too, so that problems of balancing the optical system are minimized.

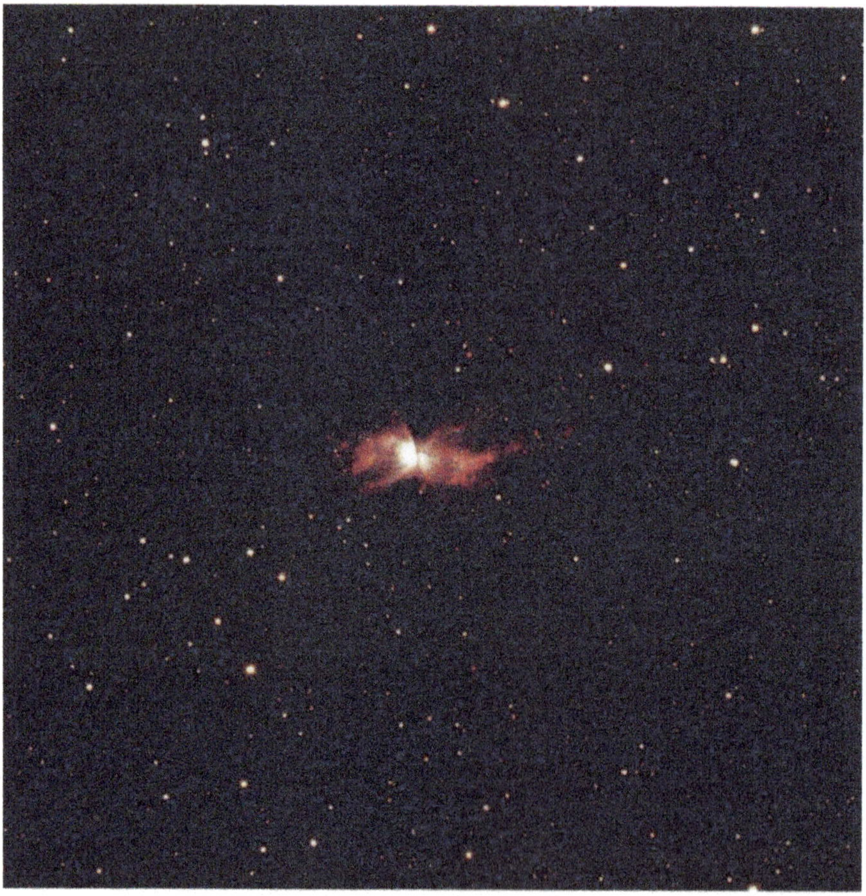

Fig. 6.9 Full RGB color combined with Hα and OIII

Filter wheels allow the user to change filters automatically during an imaging run and enable one to take composite red, green and blue exposures using an RGB filter set, or use the kind of narrowband filters examined previously. The filters are placed inside the filter wheel in a carousel, protecting them from dust and keeping them aligned. A variety of imaging software packages control the filter wheel position and automatically cycle through the filters as needed. The great thing about all this automation is that the photographer does not need to be an expert to use the filter wheel and can obtain great results in a short time. Filter wheels from different sources differ in the number of filters they can comfortably hold, but most will hold the standard filters in the UBVRI system and some additional ones such as the Hβ and the OIII filter, too.

Manufacturers of the filters discussed above usually stipulate that the filters are parfocal, that is, if placed in a filter wheel then their standard thickness ensures that their tolerances result in no need to refocus the optical system every time a new filter is used. However, this is not always the case, and as focus is a critical issue it is always best to check the system to ensure that the Airy disc—the slight ring of diffraction around a star in perfect focus—remains the same in all filters. The ability to render a faithful Airy disc every time a new filter is used is generally referred to as the critical focus zone, and this property is dependent on the f ratio of the optical system and on the wavelengths of light the photographer wishes to capture.

Although filter wheels are a bonus, even with parfocal filters it is necessary to check that the system is accurately focused every time. Experience and use will render the filter wheel and instrument parfocal in most cases, though a huge amount of setup and adjustment can be necessary if your telescope system is Portable.

Conclusion

The table below gives a ready guide to the use of the above filters in photography. Although some of the filters, such as the OIII and the LPR filters, do give the visual observer a slight advantage in seeing some objects, this cannot be applied to all filters for visual observing. The faintness of some nebulae in Hα, Hβ, OIII and SII requires long exposure imaging, and no amount of filtration will result in any improvement at the eyepiece.

Filter	Bandpass	Features
Hα	656 nm (FWHM 12 nm)	Enhanced detail in HII regions, planetary nebulae and the solar chromosphere
Hβ	486 nm (FWHM 9 nm)	Enhanced detail in HII regions, some planetary nebulae, dust lanes, dark nebulae and Bok globules
OIII	501 nm (FWHM 11 nm)	Planetary nebulae
SII	672 nm (FWHM 8 nm)	Enhances some feathery details in HII regions and extended hydrogen clouds such as NGC 7000
LPR	480 nm 520 nm 652 nm	Increases contrast in astronomical objects imaged from light polluted areas and enhances objects with HII and OIII output
Comet	501 nm (FWHM 11 nm)	Swann band filter for seeing detail in regions with CO and C_2 output such as cometary tails
UHC	486–502 nm	Enhances detail in HII regions and planetary nebulae from light-polluted areas where mercury lamps predominate, similar to some LPR filters

This rather brief roundup of commonly available astronomical filters sums up the types and applications of filters most commonly used today. Once the image is recorded with DSLR or CCD cameras the astronomer will want to progress to the next stage—reducing the photographs via *Photoshop* or similar software. A basic guide to do this can be found in the next chapter. In the final chapter of this book we shall see the nature of astronomical objects that can be revealed by using such filters.

Further Reading

Covington M (2007) Digital SLR astrophotography. Cambridge University Press, Cambridge, UK
Griffiths M (2012) Planetary Nebulae and how to observe them. Springer, New York
Mobberly M (1999) Astronomical equipment for amateurs. Springer, New York

Chapter 7

A Brief Introduction to Astrophotography and Image Manipulation

On the face of things, photographing any astronomical object seems to be a daunting task. Most of the objects are faint or small, and the average astronomers list of photographic equipment could be limited. Although this may have been true in the past, today's amateur observers have access to fantastic equipment at reasonable prices, equipment that can grab faint details and reveal deep sky objects in just a few seconds of exposure. In the last 20 years, the digital revolution has changed astronomical imaging forever. Digital SLR cameras, CCD cameras and webcams now provide the user with opportunities that previous generations could only dream of.

Why photograph the stars? Suffice to say that a photograph provides the observer with a permanent record of the observation that can be perused again and again. It also provide a visual hint to other observers who wish to track down the myriad of deep sky wonders now open to the amateur.

Camera Equipment

The advent of digital SLR cameras and CCD imaging systems means that good quality photographs are within easy reach of most amateurs. However, adding such equipment to your personal store entails a bit of spending! Good-quality DSLRs can be purchased for just a few hundred dollars, and CCD cameras can also be purchased in a similar price range, though items do obviously become more expensive dependent on quality and reliability.

This is not to say that 35-mm film has now been relegated to history. It is still a very versatile medium to work with and in expert hands provides images of

M. Griffiths, *Choosing and Using Astronomical Filters*, The Patrick Moore Practical Astronomy Series, DOI 10.1007/978-1-4939-1044-1_7,
© Springer Science+Business Media New York 2015

exquisite quality. However, film has an ethos all of its own and is a vehicle that requires some knowledge of its capabilities and drawbacks. It is hard to imagine that just a few years ago, 35-mm film was all there was available and its storage, processing and hypering were arts in themselves, without having to cope with coupling a camera to a telescope or getting a sharp image at the point where the observer was at the 'scope rather than home in the darkroom. It is best to leave the vagaries of film to one side, even though this author grew up with the method and holds a qualification in such photography.

Some starting out in astronomy will already own a digital compact camera. Rather than spending additional money on a DSLR system these compact cameras can be used at the eyepiece to photograph the Moon in some detail, and if they can be held steadily with something like a Steadypix it is possible to use the great zoom facilities on them and get a few acceptable shots of the planets, too. The exposures for such objects are relatively quick, so adjusting the exposure values (EV) down (−) and allowing the camera to adjust between the bright planet or Moon and the background sky will enable some very good shots.

Some photographers recommend such methods, as quick results can be obtained without a huge financial outlay, with the additional proviso that the most benefit can be gained even from small, undriven telescopes that many people have lingering in the attic or unused in the garage after purchasing or accepting them as gifts! Using such to photograph the Moon can transform a forgotten telescope into a valued tool and inspire the user to go further in a half-forgotten interest.

Today, the DSLR is in the ascendant, and it is becoming increasingly difficult to find good quality high ISO 35-mm film in shops, and recourse to a good photographic supplier is increasingly necessary to continue this art. For our purposes within this book, it is better to concentrate on what the market currently holds, and as the majority of cameras are now in digital format, it is these types that astronomers routinely use.

Ideally the observer should be equipped with either a CCD camera or DSLR. Choosing such equipment can be a long process of comparison and getting advice from experts in the field, but such advice is well given and worth considering so as to avoid common mistakes. A DSLR is a versatile tool, and of course is very useful for photography outside that of astrophotography, whereas a CCD camera is not.

There are many manufacturers, but the general consensus is to purchase either a Nikon, Canon or Olympus camera, as these manufacturers have a wide range of auxiliary equipment available such as lenses, T-mounts, adapters, focusing screens and filters. Although digital, these cameras follow a similar format as 35-mm film cameras in that their sizes, weights and controls are flexible, and they are easy to set and control once one has some experience with them.

It is not our intention to recommend any particular brand—excellent results can be obtained with all the above types. It is merely a choice of preference and cost. In addition to the camera, the observer will require purchasing a cable release inimical to their camera system to prevent shaking. If the observer intends to connect the camera to a telescope, then a T-mount and an adapter sleeve are also essential (Fig. 7.1).

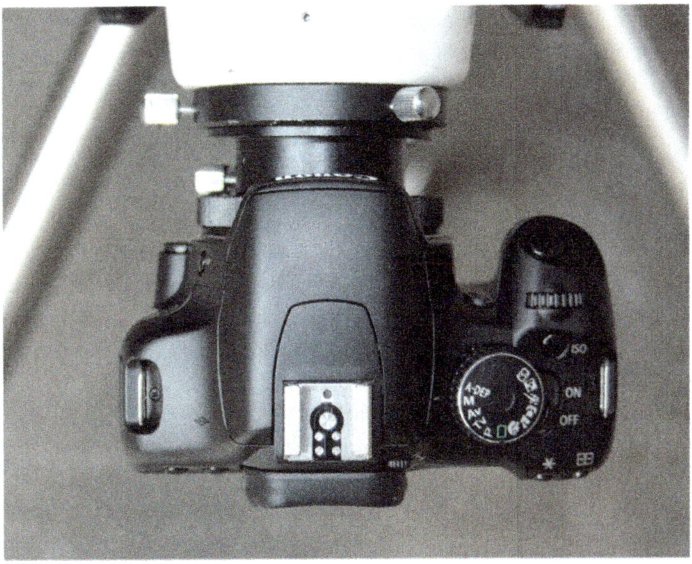

Fig. 7.1 Telescope with T-mount attached to the camera

The next problem is how to bond the camera to the telescope, or what sort of lens one is to use if the observer is mounting the camera in piggyback mode or using a driven camera mount such as an Astrotrac.

If the observer is going to use a driven camera mount, then the choice of lens will be crucial. Occasionally a DSLR can be purchased with a choice of lenses, typically a 28–80-mm focal length or a larger one with 75–300-mm focal length for example. These lenses are not built for astrophotography, though they do serve to provide wide-field shots of the sky. Naturally, any image gained will be very small, depending on the target subject and require a lot of enlargement and enhancement with this basic method. Larger examples of nebulae, such as NGC 7000 or IC 1396, or galaxies such as M31 (the Andromeda Galaxy) may be resolved with such methods and long exposures, but smaller objects will be invisible or just splotches in the background.

The best approach to photographing these objects is to connect the camera to a telescope at its focus point. This procedure then gives one the advantage of having a large telephoto lens with an f ratio exactly the same as the telescope. With this arrangement the observer doesn't have to do any complicated mathematics in working out f-ratios or thinking about exposure times. With DSLR's several photographs may be taken of the one object and then stacked in the appropriate software to produce a single, higher resolution image.

Alternately, the observer can purchase a piece of equipment commonly called a Steadypix. This functions as a connector between the camera lens of your DSLR and an eyepiece of choice and fixes to any eyepiece on the eyepiece holder. Such photography is called *eyepiece projection* or *afocal astrophotography*.

If one is using this arrangement, remember to put the camera lens at its lowest f-stop (f2.8 or 5.6) and focus it on infinity (by placing it on manual focus rather than automatic!). Then the focal ratio must be worked out, which can be done with the following equation:

$$\mathrm{Fr} = \left[\left(\mathrm{C}/\mathrm{E}\right) - 1\right]\mathrm{T}$$

where Fr is the focal ratio, C is the focal length of the camera lens, E is the focal length of the eyepiece and T is the focal ratio of the telescope. To find the focal ratio of your telescope, divide the instrument aperture into the focal length. If you have a 1,000-mm focal length telescope with an aperture of 200 mm, then the focal ratio will be 1,000/200 = f5.

In the above configuration a telescope of focal ratio f10 using a 50-mm camera lens and a 25-mm eyepiece will have the sum 50/25 mm − 1 = 1. This single unit can then be multiplied by the telescope's focal length (f10) to get the focal ratio of the entire system. With an eyepiece of 15-mm focal length, the resultant focal ratio with this telescope will be f23 (50/15 mm − 1 × T) and so on as one uses different eyepieces.

Although occasionally afocal astrophotography enables a relatively large initial image to be obtained, it has drawbacks in that the system must be accurately driven, the contrast of the system will decrease with increasing magnification and some large focal ratios of f40 and above are just giving diminishing returns. Additionally, focusing the system accurately can become very problematic. It is best to attach the camera directly to the telescope with a T-mount and play with the image software later, though many amateurs have gained good results from using Steadypix systems. Again, it is a question of preference.

Alternatively, many astronomers have CCD cameras such as the various Atik cameras, models by the Santa Barbara Instruments Group (SBIG) or Starlight Express. Many of the models are full-color CCDs; others are black and white and require colored filters such as RGB or BVR to produce a full-color image. Once again, choice is down to observer preference and costs. The advantage of the CCD camera over the DSLR is the smaller field of view and the rapidity of capture and quality of the images, which can be manipulated in various programs such as *Maxim DL*, *Artemis* and others.

Webcams specifically made for astronomy may also be employed in a similar way to CCDs, although the resolution depends on the chip. One of the most popular of such items is the DMK camera, which requires a program entitled *Registax* (or similar) to complete the imaging processing. If you wish to use this form of photography, see Massey and Quirk's *Deep-Sky Video Astronomy* (Springer, 2009), which covers in depth the video and *Registax* techniques needed to take great pictures.

There are so many variants of CCD camera available that it is not possible to cover them all in great detail, and it is recommended that the observer read the reviews of such products online to make the best choice.

Resolution of the CCD camera or any DSLR is the most important thing for an amateur astronomer. For the observer equipped with or contemplating purchasing a CCD camera, the chip size and pixel numbers on the axis of the chip are very important in obtaining detail and ensuring that a CCD camera and telescope will resolve any astronomical object sufficiently. Thankfully most astronomical bodies are quite large in comparison to the pixel size of a CCD, but with any camera you consider purchasing use this simple equation, which will enable you to ensure that the CCD camera of your choice is adequate.

The relationship is:

$$R'' = \left[\left(Ps_m 206\right) / a\right] / f$$

where R'' is the angular resolution in arc seconds, Ps_m is the pixel size, a (in mm) is the telescope aperture and f is the focal ratio of the telescope. If the camera has pixels of 10-m size and is fitted to a 200-mm aperture f5 telescope, then:

$$10 \times 206 = 2,060$$

$$2,060 \div 200(a) = 10.3$$

$$10.3 \div f5 = 2.06$$

The chip therefore has a resolution of 2.06″.

If you then need to know the size of the field of view through such a camera the simple relationship is:

$$S^\circ = (Pr\,Pn) / 3,600$$

where S^o is the size of the frame in degrees, Pr is the pixel resolution and Pn is the number of pixels along the axis of the chip divided by 3,600. So for a chip with 1,500 pixels along its main axis:

$$1,500 \times 2.06 = 3,090$$

$$3,090 \div 3,600 = 0.85^\circ$$

This is larger than the full Moon and will encompass most astronomical objects, but obviously smaller, more distant objects will lose some resolution despite the fact that the image can be enlarged.

No matter what the individual uses, one of the factors essential to any photography is the focus; it must be sharp and free of any obvious defects such as smearing at the edges. It may take time and practice for any astronomer to achieve a good focus, but it is well worth the effort, as anything that is out of focus is disappointing when one looks at the pictures. Focusing is a slightly frustrating task when out in the field, yet is the most essential component of any photography, and getting the focus right deserves time and attention. Many DSLRs have a 'live-view' feature, a consideration to be taken into account when purchasing one, while a CCD has the ability to download a picture there and then to judge for focus and quality.

For the UBVRI filters that are commonly used with CCD cameras it is recommended that the reader refer to the last chapter for the typical bandwidths, filter systems and filter wheels to use with these cameras.

"How-to" Books and Software

Once the observer has this equipment and a little experience in using it, most will just get on with imaging astronomical delights without a care in the world. Experienced observers using such equipment perform the usual controls of taking dark and flat fields to maximize the image on the CCD along with windowing, bias framing, binning and maximizing the gain. CCD imaging is an art in itself and readers are advised to ask advice of practiced astrophotographers or read through any of the books on the subject to gain expertise and skill.

There are several excellent books on reducing astronomical images from CCDs and DSLRs. Suffice to say that if the reader requires more information and depth then please access such books as Steven Seip's *Digital Astrophotography* (Rocky Nook Publishing, Santa Barbara, 2009), David Ratledge's *Digital Astrophotography – The State of the Art* (Springer, New York, 2005), or Ireland's massive work *Photoshop Astronomy* (Willmann-Bell, 2009). These provide a varied and deep introduction to the subject and complement each other well.

In the past a number of steady and accessible guides have appeared such as Michael Covington's *Astrophotography for the Amateur* or *Digital SLR Astrophotography* (Cambridge University Press, 2007); Jerry Lodriguss' *A Guide to Astrophotography with Digital SLR Cameras* (Astropix, New Jersey, 2007); Steve Howell's *Handbook of CCD Astronomy* (Cambridge University Press, 2006), or Greg Parker's *Making Beautiful Images: Astrophotography with Affordable Equipment and Software* (Springer, New York, 2007), just to name some examples. The British Astronomical Association sells a small booklet for novice astrophotographers entitled *Introduction to DSLR Astrophotography,* and other national societies worldwide may do the same.

If one is using a digital camera of any type, it is necessary to use support software that either comes with the camera or is compatible with the camera format. Canon and Nikon both produce their own reduction software that is fairly basic, and observers will move away from these as quickly as possible, as they have very limited capabilities. Most astrophotographers use *Photoshop* and its variants such as *Lightroom*, or *Paintshop Pro*, *Maxim DL, GIMP* and others to manipulate their images. Some of these programs are industry standard, and it is likely that the observer has come across them in everyday life and is relatively familiar with their use. If not, then learning image manipulation is another skill that has uses beyond the field of astronomy; in today's job market, the ability to use many software packages is something that many employers look for.

The essential feature that the software should have is the ability to correct the colors, background, shadowing and light levels of the image. The ability to stack several images together or to stitch a mosaic is also necessary, and these skills can be learned quickly merely by playing with the images. Don't be afraid of making mistakes; in the world of software they can always be undone! In *Photoshop* most of the features stated here can be found under the single menu heading 'enhance,' and other programs have similar transferable menu headings that enable the user to substitute quickly between programs. *Photoshop* is probably one of the best programs available, and it or its derivatives such as *Photoshop Elements* can be purchased for a nominal sum.

With a CCD or webcam, however, *Photoshop* manipulation probably comes last, as programs such as *Maxim DL* or *GIMP* provide their own reduction techniques by taking the images as Flexible Image Transport System (FITS) files that can be altered in these programs initially or by the software *FITS Liberator* (available as a free download) before being transferred to *Photoshop* or equivalent if necessary. The raw files from a DSLR camera can be processed in a program such as *Deep Sky Stacker*, which is available as a free download and is useful in that it adds your dark frames, bias and flat frames to your images and stacks the selected frames automatically so you can get on with other things (like sleep!) or transfer the finished files across (in TIFF format) to *Photoshop* if necessary. Most experienced astrophotographers use such software as a general photo editor before moving the files across to the software equivalents of their big brothers.

Just as in the purchase of a camera, the use and application of software to reduce and alter images is a question of choice in most cases. Nevertheless such software is an essential part of getting the best from your shots. The amount of information that a CCD or DSLR chip records is amazing; manipulating shots brings out the maximum from the hidden data and can turn a seemingly mediocre picture into a work of art. Ultimately, it is up to the individual to make the most of the picture with the best-quality software available.

The use of filters in imaging has inevitably led to the growth of software packages that process the files from DSLR or CCD cameras. Classes teaching the basics of these software items can be found almost anywhere on a local level, though most astronomers use these packages from a self-taught perspective. Software programs such as *Paint Shop Pro* and *Adobe Photoshop* are now commonplace in the industry and are sophisticated programs that require some training to use to its fullest. It is not our intention here, though, to provide that training. We simply wish to introduce *Photoshop* in a way in which everyone can use it right from the start.

Preparing the Image in *Photoshop*

Most astronomers and photographers will have some access to *Photoshop*. One of the current packages for the professional is *CS 6*, and this is complemented by a slightly "simpler" package called *Photoshop Elements*. *Photoshop CS 6* also has access to astronomy plug-ins that will make your imaging take on a life of its own.

Photoshop has one flaw however—it cannot accept the standard format files of the raw images from both DSLRs and CCD cameras. These files are known as the Flexible Image Transport System, or FITS files, which need an additional program such as *Maxim DL* or the widely available *FITS Liberator 3* (which can be downloaded free from the Internet at http://www.spacetelescope.org/projects/fits_liberator/).

Much processing of the image can be done in *Maxim DL* alone, but as many have access to *Photoshop,* we will feature this package and take the reader through a basic exercise to get a good color image using *FITS Liberator,* though we will include the basics for *Maxim DL*, too, as the exercise for each is very similar for reduction to *Photoshop.*

The exercise below assumes that you have taken the images in RGB (BVR) with a monochromatic CCD camera. A digital SLR camera can be placed on a monochrome setting, and use of filters will then follow exactly what a CCD camera achieves, though some experience in this field is preferable.

FITS Liberator and *Maxim DL*

The object here is to prepare your images for *Photoshop*, so follow these simple instructions to achieve this.

Firstly, download your FITS files (usually named as Raw) into a folder from the DSLR—if you are using a CCD they should already be saved in a folder. Open *FITS Liberator* and open your **blue.fits** image for the object that you have photographed. In the figures that appear below we have used the author's images of NGC 5128 from the LCOGT telescope network.

You will see that there is a function on the right hand side of the picture window called the **Stretch Function**. You can either click **auto** scaling or play with the sliders under the picture window until you are happy with the quality of the image. When you have done this go to **channels** under **auto** scaling and select the **8-bit** function there. Once this is done, go to **save and edit** and save your file as **NGC 5128 Blue.tiff.** Repeat this operation for the red and green fits files. See the *FITS Liberator* window here in Fig. 7.2 for comparison.

In Maxim DL the above functions can be found by opening your blue FITS file and going to **view** and **screen stretch window.** Leave the setting there on **manual** and center the object in the picture window before using the **minimum or maximum** functions. Again, play with the sliders until you are happy with the quality of the image. *Maxim DL* allows you to save the image as a JPEG file, which is the usual format for *Photoshop.* Save your images as JPEG files and then open *Photoshop.* One can see the *Maxim DL* screen shot in Fig. 7.3.

For both the files processed from *Maxim DL* (although as stated previously *Maxim* will process the images for you without using *Photoshop* if you know how to use the package) and *FITS Liberator* the same pathway through *Photoshop* is now open to the reader.

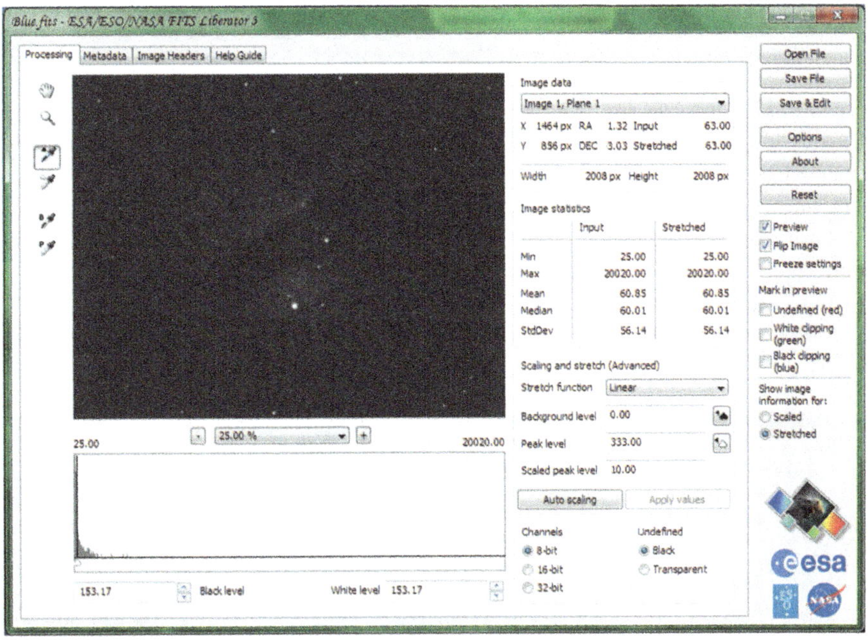

Fig. 7.2 *FITS Liberator* screen

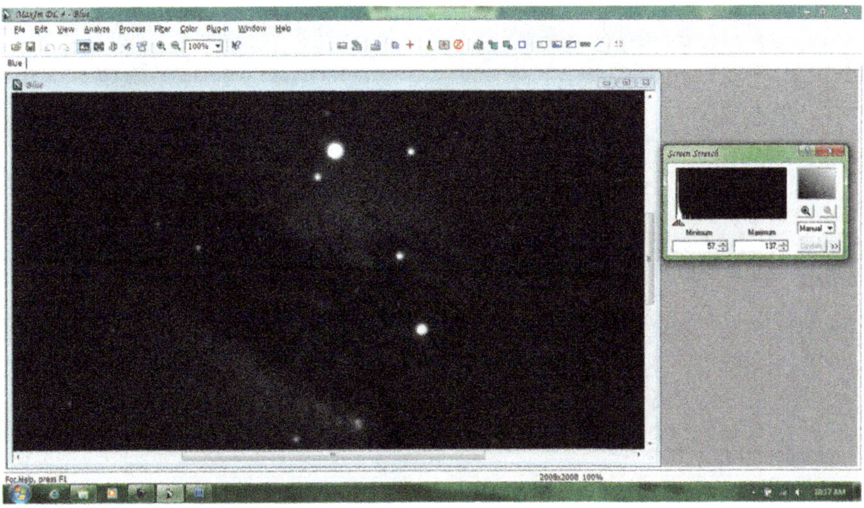

Fig. 7.3 *Maxim DL* screen shot

Photoshop and Processing Your Image

Open *Photoshop* (we are using *Photoshop Elements* here for the screenshots rather than *CS 6*, although most of the dropdown menus will be similar), and then open your RED.TIFF or RED.JPEG file, depending on your use of *FITS Liberator* or *Maxim DL*.

Once the picture is open, go to **image** and then **mode** and select the **RGB color** item on the dropdown menu, as can be seen in Fig. 7.4.

Once this is done you may go to **enhance** (image>adjustment in *CS 6*) and pick **adjust color**, which will bring up a new menu box. Pick **adjust hue/**

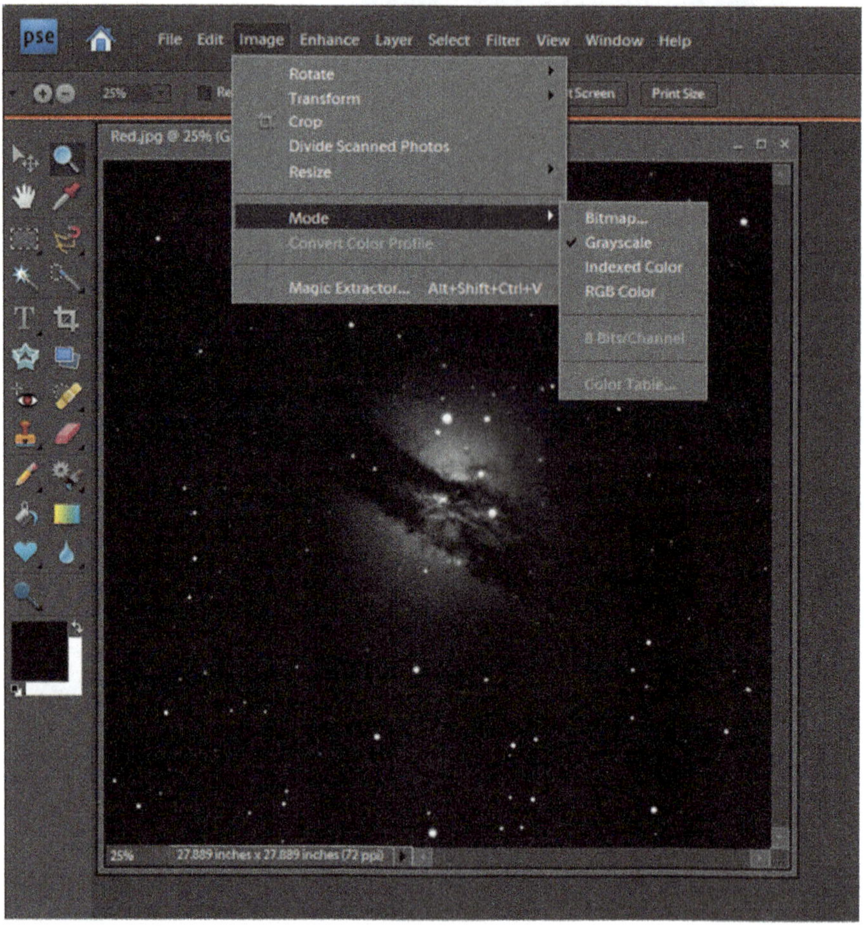

Fig. 7.4 Processing RGB mode

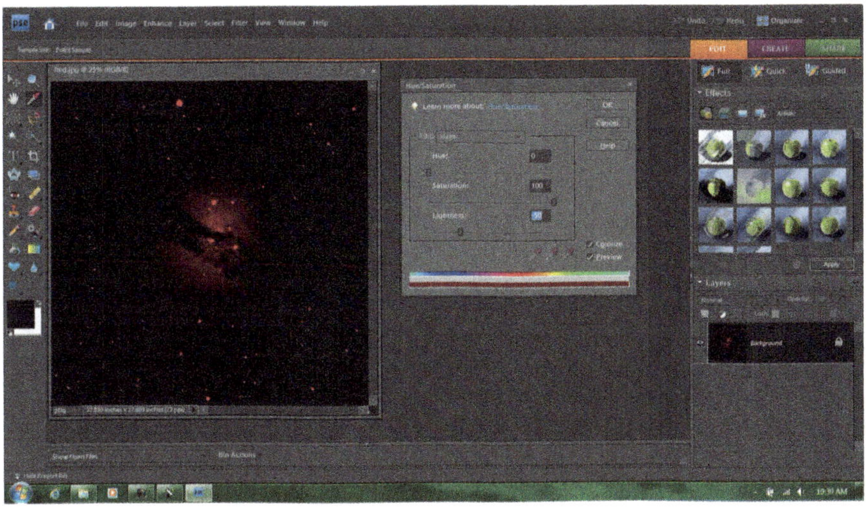

Fig. 7.5 Colorize the image using hue and saturation

saturation from this box, and it will open a new palate for you. Click the **colorize** box on the lower right hand side and then adjust the hue and saturation, as can be seen from Fig. 7.5.

You then have to follow the same steps for your green (V band) image and your blue one, but the amount of colorization will depend on several factors. The table below is a rough guide to what levels you will need to take each image to, to achieve a good overall color contrast in the final image.

Color	Hue	Saturation	Lightness
Red	0	100	−50
Green	120	100	−50
Blue	240	100	−50

Once you have all three red, green and blue images colorized to the right degree the reader will need to select the green image by using **ctrl-A** to select the entire image then **ctrl-c** to copy the image. Once this is done select the red image and overlay the green copied image by selecting **ctrl-V**. This should not only overlay the green image on the red one but also bring up a new layer on the **layers palette** on the bottom right hand side of the screen. Do the same exercise for the blue image and overlay it on the new red-green layer until you have three layers in the layers palette.

Below the layers palette header is a button with **normal** on it. This is your **blending modes menu**. Click on this and a menu should come up. You then need to select **screen** as can be seen in Fig. 7.6.

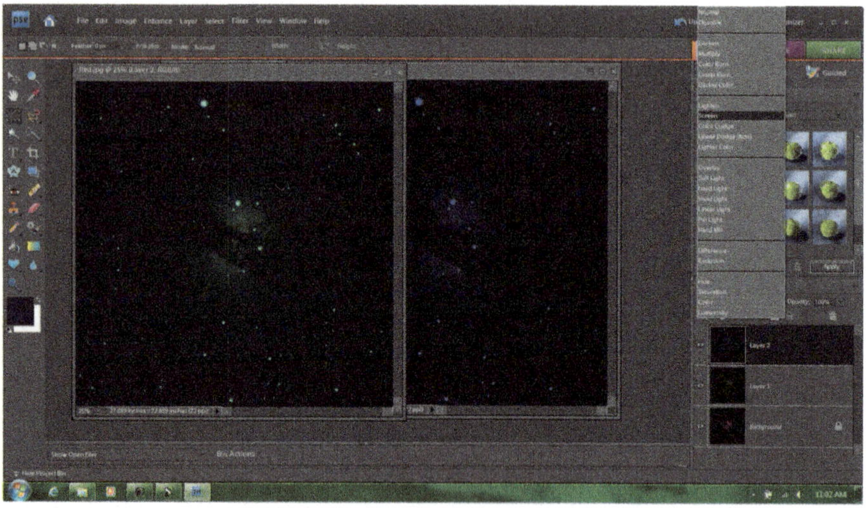

Fig. 7.6 Layers palette

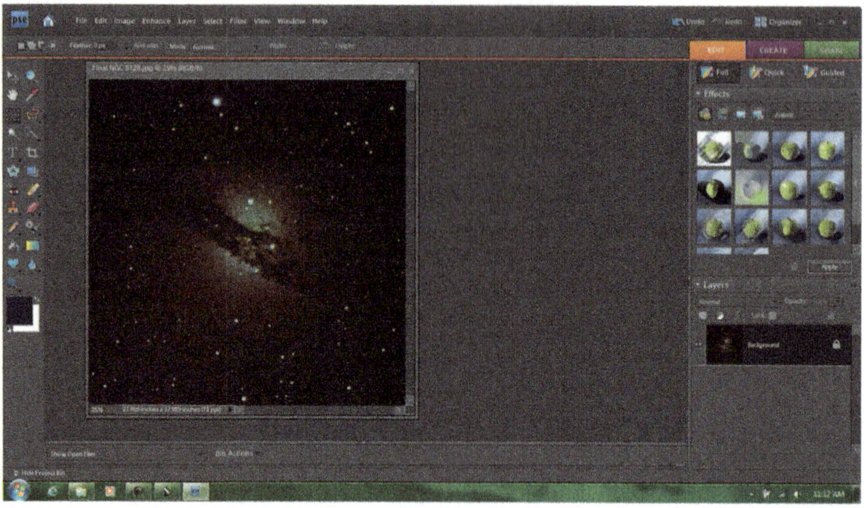

Fig. 7.7 Final saved image

What the screen selection does is overlay each color but lets you see the image colors as they should finally be. You will not be able to make any further adjustment to the image until you go to the **Layer** button at the top of the screen and pick the **flatten image** button on the dropdown menu. This then prepares the final image for adjustment. Select **Enhance** (image>adjustment in *CS 6*) and **adjust lighting>levels** and then you can enhance the image by selecting the **channel** button and manipulating the individual colors and brightness as required as can be seen in Fig. 7.7. Once you have your color image and you are happy with it then you can save it.

That is basically the exercise that the reader will have to perform each time and is very familiar to all photographers that use this system. With additional narrowband filters such as Hα, OIII or SII the addition of a *luminance layer* is required.

Luminance Layers

You need to perform exactly the same steps with a narrowband image as we have seen above as far as selecting it as an RGB image. You will not be colorizing it but you will be overlaying it in much the same way as above. Once you have your final image prepared from your original RGB images you need to select your Hα layer (after going through image>mode>RGB color) and saving it as a Hα layer. With your color image and the Hα image on the screen select the layers palette and select **luminosity** from the menu that drops down under the **normal** button.

There may be colored haloes around the stars in the original color image, so remove these by using the **healing brush tool** and remove the stars from the color image if this is an obvious problem. You will have to do this individually for every star. Although this may be long, drawn out process it brings out great detail in your final images once you have flattened all the images together.

Registax

Registax is free astronomy image and video manipulation software that is available as a windows only product at www.astronomie.be/registax/. It produces astronomical images by taking all the image data from a webcam exposure and stacking the photographs so that one high-resolution image is obtained. *Registax* enables the image to be free of atmospheric shimmering and bad seeing conditions that spoil the image of a single-shot camera. It has several versions, but at the time of writing the latest release is *Registax 6.* There are similar packages available online, but this author has used this excellent program on many occasions and uses *Registax 5.1* as a matter of preference.

Registax takes the image files from a DMK or webcam in the form of an .AVI file. Open the file by going to **Select** and then clicking on the AVI file saved from the camera. This will bring up the following panel, as seen in Fig. 7.8. The reader then needs to select the **multi-align** on the image by clicking on the circle in the **alignment method box.** Also, click **32** on the **align box size** so that fine features can be selected across the image area. Once this is done, click selected areas on your open image and, once happy with the selection, click **Align,** which has a green line under it.

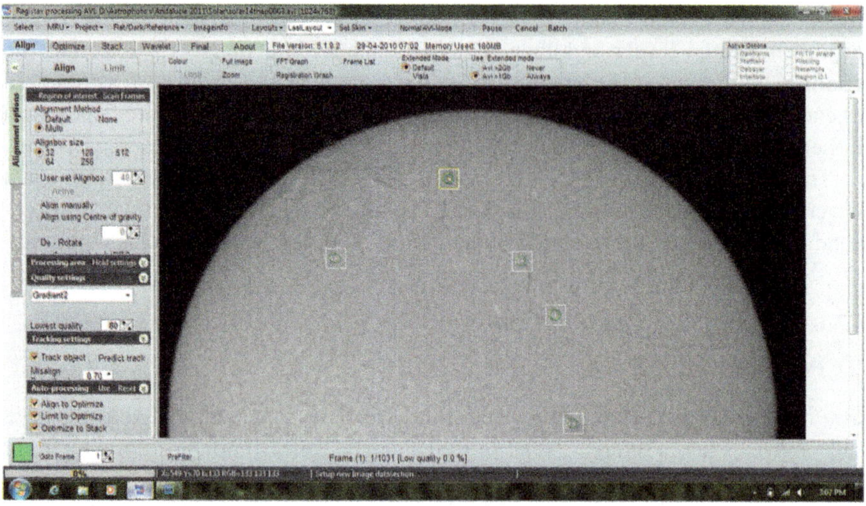

Fig. 7.8 *Registax* front page

At the bottom of the page you will see an increasing percentage—up to 100 % of how the image is being aligned. *Registax* is a very versatile tool, as from this stage on the software does all the work for you and you merely have to follow its suggestions, to a point, which are indicated by the placement of the green line under align and Limit.

The next stage will be to optimize the image, so click on the **Optimize** tab next to the Align tab. Select **Optimize and Stack,** and the screen will come up with small images of the multi-highlighted sections that you selected from the align tab. Again, the left hand bottom of the page will show the user an increasing percentage until 100 % is reached and the image is finalized. Once the increment reaches 100 % the page will automatically go to **Stack,** and the images will be stacked for use. Depending on how many images were taken and the size of the AVI file, this could take several minutes, but again will be incrementally read out at the foot of the page. This process can be seen from Fig. 7.9.

Once the stacking has been achieved, the software will automatically move to **Wavelet** for adjusting the image. This is where the sliders on the left of the page can be used, but beware! Use them sparingly to bring out the fine details within your image, and remember that these fine details may look great in *Registax* but can come over with too much contrast once the final image is taken into *Photoshop.* Adjust the image accordingly using the wavelets and then observe the image. Not all has yet been adjusted, so the user needs to click the underlined button **Do All**. Once this is done simply save the image as a JPEG file, which is the default setting.

From here one can then open the image in *Photoshop* and go to **enhance > adjust color > hue and saturation.** After clicking the **colorize** button on the bottom

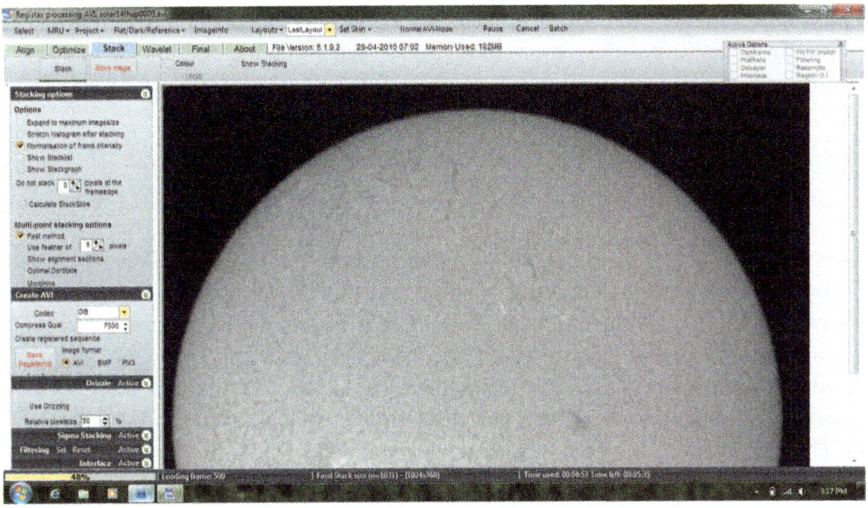

Fig. 7.9 Stacking

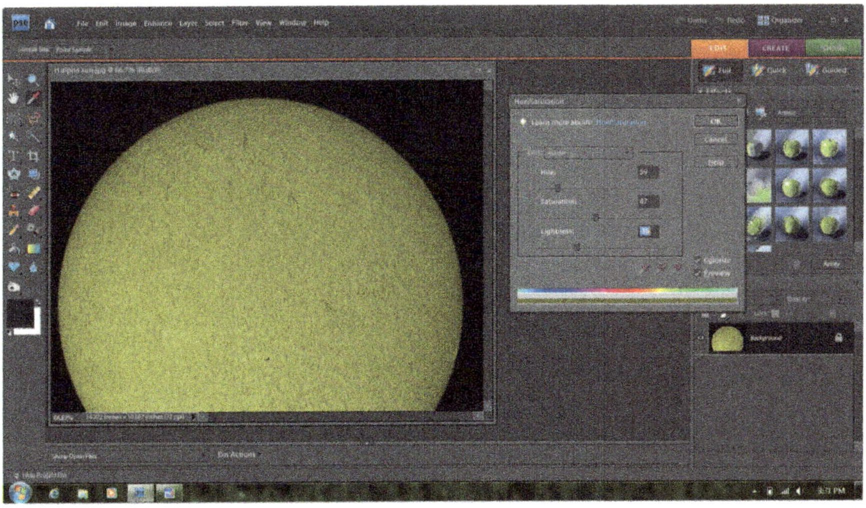

Fig. 7.10 Adjusting the color

right of the box and adjusting the sliders to the correct color then the image can be finalized as can be seen here in Fig. 7.10 as a final single-shot image of the video taken by the webcam.

The versatility of DMK cameras and webcams is wonderfully demonstrated by imaging the Sun, Moon and planets. Coupled to powerful programs such as

Photoshop and *Registax,* even a small telescope can obtain fantastic images in relatively short exposures, even though the usual monochromatic image may require additional manipulation.

Conclusion

That basically is how to obtain colored images from the raw FITS files from your DMK or CCD camera. Obviously this exercise is rather simplistic, but if it is followed strictly then good results can be achieved. Experienced observers and photographers will probably find a myriad of ways of accomplishing this task and steadily improve on their basic techniques, using the add-ons and plug-ins that come with various packages and make astroimaging such a worthwhile task.

The next task of course is to select astronomical objects worthy of study. The next chapter will take the reader through over 100 objects that cover the personal favorites of many astronomers. Enjoy the selection! However, there are many others that can be added to this list.

Further Reading

Gendler R (2013) Lessons from the masters: current concepts in image processing. Springer, New York

Ireland RS (2009) Photoshop astronomy. Willmann-Bell, Richmond, USA

Mobberley M (2006) Lunar and planetary webcam users guide. Springer, New York

Chapter 8

Observing and Imaging Objects

The whole point of using filters in astrophotography is to bring out the maximum amount of detail in the object being imaged and presenting as faithful a representation of that object as possible. Beyond this aesthetic concern the use of filters in imaging enables professional astronomers to gain information from astronomical bodies at specific wavelengths that render important insights into the nature and physical processes in the objects.

Nevertheless beyond the scientific application of filters there lies a beauty and wonder in astronomical objects that continue to enhance everyone's appreciation for the awesome universe that we inhabit. The pictures returned by the Hubble telescope have been featured on the front pages of many newspapers worldwide, and there is hardly a day that goes past when some feature of astronomical news does not make it into the media. The abilities of modern technology have enabled the ordinary amateur astronomer to capture the charm and subtleties of many astronomical objects, and the pursuit of astrophotography has grown from a small-scale interest to multi-million dollar businesses.

The following selection of objects represents some of the brightest and more interesting items available to both northern and southern hemisphere observers. With every object we have recommended a narrowband filter to use in addition to the ubiquitous RGB (or UBVRI) filters for correct color rendition, as discussed in previous chapters. The objects represent a wide variety selected because of their astrophysical importance and their artistic beauty. No doubt the reader can add many more to the list here.

The catalog list below is a quick reference guide to the objects in right ascension order, which is then followed by individual objects and a brief description of each with some pertinent details. Although there are recommendations for the best filters to use with each object, seasoned photographers will no doubt try a combination of

M. Griffiths, *Choosing and Using Astronomical Filters*, The Patrick Moore Practical Astronomy Series, DOI 10.1007/978-1-4939-1044-1_8,
© Springer Science+Business Media New York 2015

additional filters to bring out intimate details inimical to the filter used. Some objects will respond well to narrowband filters, while some will probably not need more than RGB filters to get a beautiful shot. Choice will be down to the individual and the camera/telescope set up. No matter what is used, it is hoped that this selection meets with the reader's approval and encourages the observer to try imaging these astronomical treasures.

Object	Con	RA	Dec	Mag	Type
NGC 104	Tuc	00h 24m	−72.04	4.9	GC
NGC 246	Cet	00h 47m	−11.52	8.2	PN
NGC 253	Scl	00h 47m	−25.17	8	GAL
NGC 281	Cas	00h 52m	+56.37	12.8	NEB
NGC 598/M33	Tri	01h 33m	+30.39	5.7	GAL
NGC 628/M74	Pis	01h 36m	+15.47	10.2	GAL
NGC 650/M76	Per	01h 42m	+51.34	10.1	PN
NGC 891	And	02h 22m	+42.2	10.8	GAL
NGC 1300	Eri	03h 19m	−19.24	11.4	GAL
IC 1805	Per	02h 32m	+61.29	6.8	NEB
NGC 1333	Per	03h 29m	+31.20	6.5	NEB
NGC 1360	For	03h 33m	−25.52	9.4	PN
M 45	Tau	03h 47m	+24.07	2	OC
NGC 1491	Per	04h 03m	+51.18	8.9	NEB
NGC 1499	Per	04h 03m	+36.25	6.4	NEB
NGC 1501	Per	04h 06m	+59.54	10.2	PN
NGC 1514	Tau	04h 09m	+30.46	9.4	PN
NGC 1535	Eri	04h 14m	−12.44	9.2	PN
NGC 1554	Tau	04h 21m	+19.32	12.3	NEB
IC 2118	Eri	05h 02m	−7.54	13.1	NEB
NGC 1788	Ori	05h 06m	-3.21	9.2	NEB
IC 410	Aur	05h 22m	+33.23	11.1	NEB
NGC 1931	Aur	05h 31m	+34.15	10.1	NEB
NGC 1952	Tau	05h 34m	+22	8.4	SNR
NGC 1975/M42	Ori	05h 35m	−05.23	4	NEB
NGC 1977	Ori	05h 35m	−4.48	7.1	NEB
Sim 147	Tau/Aur	05h 41m	+27.46	12.5	SNR
NGC 2024	Ori	05h 41m	−1.51	5.6	NEB
IC 434	Ori	05h 41m	-2.3	7.3	NEB
NGC 2068/M78	Ori	05h 46m	+0.03	8.3	NEB
NGC 2070	Dor	05h 38m	−69.05	7.6	NEB
NGC 2170	Mon	06h 07m	−06.23	9.4	NEB
IC 443	Gem	06h 17m	+22.31	11.9	NEB
NGC 2237	Mon	06h 33m	+4.59	9.1	NEB
NGC 2261	Mon	06h 39m	+0.45	9	NEB
NGC 2264	Mon	06h 41m	+9.53	4.3	NEB
NGC 2362	Can Maj	07h 18m	−24.57	4.1	OC
NGC 2392	Gem	07h 29m	+20.54	10.2	PN

(continued)

(continued)

Object	Con	RA	Dec	Mag	Type
NGC 2437/8	Pup	07h 41m	–14.49	6.1	OC/PN
NGC 2403	Cam	07h 36m	+65.36	8.9	GAL
NGC 2440	Pup	07h 41m	–18.12	9.4	PN
IC 1396	Cep	08h 57m	+54.57	10	NEB
NGC 2818	Pyx	09h 16m	–36.37	5.6	OC/PN
NGC 2841	Ur. Maj	09h 22m	+50.58	10.1	GAL
NGC 2903	Leo	09h 32m	+21.3	8.4	GAL
NGC 3034/M82	Ur. Maj	09h 55m	+69.4	8.4	GAL
NGC 3132	Vel	10h 07m	–40.26	9.8	PN
NGC 3195	Cham	10h 09m	–80.51	11.6	PN
NGC 3242	Hyd	10h 24m	–18.38	8.6	PN
NGC 3372	Car	10h 45m	–59.52	2.8	NEB
NGC 3587	Ur. Maj	11h 14m	+55.01	9.9	PN
NGC 3628	Leo	11h 20m	+13.35	9.9	GAL
NGC 4258	Can Ven	12h 18m	+47.18	9.1	GAL
NGC 4321	C. Ber	12h 22m	+15.49	10.1	GAL
NGC 4361	Cor	12h 24m	–18.47	10.9	PN
NGC 4485	Can Ven	12h 30m	+41.41	11.7	GAL
NGC 4486	Vir	12h 30m	+12.23	9.5	GAL
NGC 4565	C. Ber	12h 36m	+25.59	10.4	GAL
NGC 4594	Vir	12h 39m	–11.37	8.9	GAL
NGC 4631	Can Ven	12h 42m	+32.32	9.8	GAL
NGC 4656	Can Ven	12h 43m	+32.1	11.1	GAL
NGC 4755	Crux	12h 53m	–60.22	4.1	OC
NGC 5053	C. Ber	13h 16m	+17.42	9.4	GC
NGC 5128	Cen	13h 25m	–43.01	6.84	GAL
NGC 5139	Can Ven	13h 26m	–47.28	3.7	GAL
NGC 5189	Musca	13h 33m	–65.58	10.2	PN
NGC 5194/M51	Can Ven	13h 29m	+47.11	8.4	GAL
NGC 5236	Hyd	13h 37m	–29.51	7.5	GAL
NGC 5457	Ur. Maj	14h 03m	+54.2	8.3	GAL
NGC 5907	Dra	15h 15m	+56.19	11	GAL
NGC 6121/M4	Sco	16h 23m	–26.31	5.8	GC
IC 4604	Oph	16h 25m	–23.27	11.4	NEB
NGC 6188	Ara	16h 40m	–48.47	8.0	NEB
NGC 6302	Sco	17h 13m	–37.06	10.3	PN
NGC 6337	Sco	17h 22m	–38.29	12.6	PN
NGC 6514/M20	Sag	18h 02m	–23.01	6.1	NEB
NGC 6523/M8	Sag	18h 03m	–24.23	4.9	NEB
NGC 6572	Oph	18h 12m	+06.51	9.2	PN
NGC 6611/M16	Ser	18h 18m	–13.49	5.8	NEB
NGC 6618/M17	Sag	18h 20m	–16.1	5.8	PN
NGC 6720/M57	Lyr	18h 53m	+33.01	8.8	PN
NGC 6781	Aql	19h 18m	+6.32	11.4	PN

(continued)

(continued)

Object	Con	RA	Dec	Mag	Type
NGC 6818	Sag	19h 43m	−14.09	8.4	PN
NGC 6822	Sag	18h 44m	−14.47	9.3	GAL
NGC 6826	Cyg	19h 44m	+50.31	8.8	PN
NGC 685/M27	Vul	19h 59m	+22.43	7.5	PN
NGC 6888	Cyg	20h 12m	+38.21	7.6	NEB
NGC 6960	Cyg	20h 45m	+30.42	7.1	SNR
NGC 6905	Vul	20h 22m	+20.06	11.1	PN
NGC 6992	Cyg	20h 45m	+30.42	7.1	SNR
IC 5067	Cyg	20h 52m	+44.25	4.3	NEB
NGC 7000	Cyg	20h 59m	+44.31	4.3	NEB
NGC 7008	Cyg	21h 00m	+54.32	10.7	PN
IC 5146	Cyg	21h 53m	+47.16	12.1	NEB
NGC 7009	Aqu	21h 04m	−11.21	7.9	PN
NGC 7023	Cep	21h 01m	+68.10	6.8	NEB
NGC 7293	Aqu	22h 29m	−20.58	7.1	PN
NGC 7635	Cas	23h 20m	+61.12	10.2	NEB
NGC 7662	And	23h 25m	+42.32	8.6	PN
NGC 7822	Cep	23h 57m	+68.22	10.8	NEB

NGC 104 Tucana

RA: 00h 24m
Dec: −72.04
Size: 30.9′
Magnitude: 4.9
Filter: RGB

NGC 104, or 47 Tucanae, as it is commonly known, is one of the most wonderful globular clusters in the sky. Too far south for northern observers, this deep-sky jewel is an enormous aggregation of stars lying 16,700 light years away, with a diameter of 120 light years and containing over one million stars. Naturally orange-red in color due to the abundance of old dwarf and red giant stars it contains, this fantastic object is easily visible, and many southern observers make it a point to image this cluster as their introduction to astrophotography. Common RGB filters build up a lovely picture of this excellent object after just a few seconds exposure. Its rich center can become burned out with overexposure, so this is an object that will need careful integration in *Photoshop* of long and short exposures to bring out all the detail. (Image taken by Hartpury College with Faulkes Telescope South operated by Las Cumbres Observatory Global Telescope Network. Used with permission.)

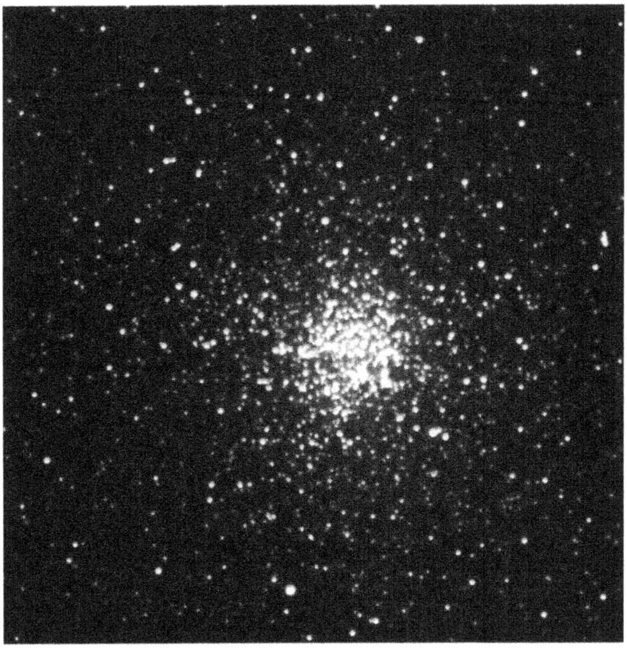

Fig. 8.1 NGC 104

NGC 246 Cetus

RA: 00h 47m 03s
Dec: −11.52.3
Size: 4′
Magnitude: 8.4
Filters: Hα, OIII, SII, RGB

Visually this is a diffuse, white glow in the body of Cetus about 4′ across. Slightly structured into arcs, with the SW side the slightly brighter part of the nebula, it has three stars clearly visible on the body, with a central one glowing white in a gray cast of nebulosity. Not a difficult object to image if sky conditions are good, though it can get lost in the murk along the horizon from the UK. Use of Hα and OIII filters are recommended to bring out the entire body in addition to standard RGB, as the nebulae is diffuse yet intricately structured. NGC 246 is also one of the Herschel 400 objects and has the common name of the "Skull Nebula." (Image taken by Sarah Roberts with Faulkes Telescope South operated by Las Cumbres Observatory Global Telescope Network. Used with permission.)

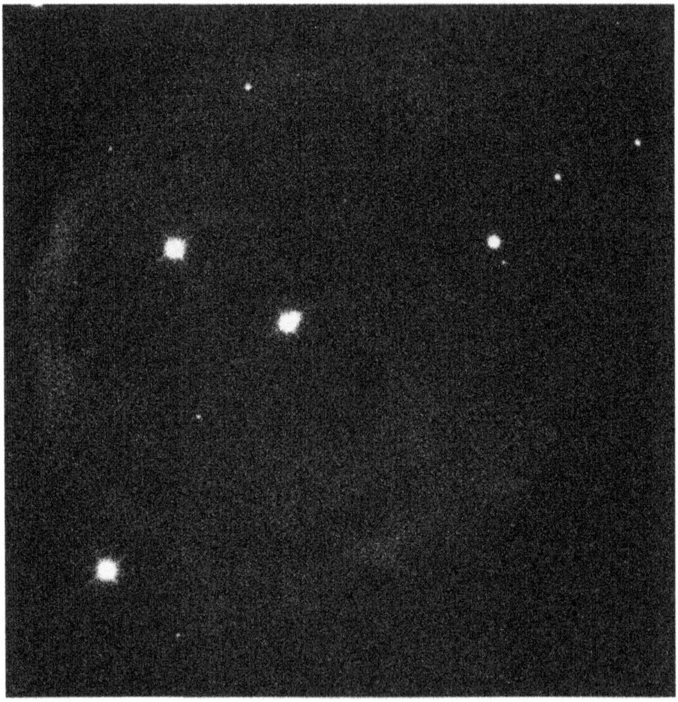

Fig. 8.2 NGC 246

NGC 253 Scutum

RA: 00h 47m
Dec: −25.17
Size: 27×7′
Magnitude: 8.1
Filters: Hα, RGB

This is the superb "Silver Coin Galaxy," which has a great range of detail requiring careful exposure in RGB and Hα to bring out the wonders of the spiral arms, which glow with an electric blue light from countless O- and B-type stars recently formed during a "starburst" phase and the delicate HII regions that mottle the body of the galaxy. Lying at just over 11 million light years away, the galaxy exhibits great detail, and the arms and dust lanes are clearly evident in short exposures. Given its large size, it may be necessary, depending on the kind of equipment in use during imaging, to take several shots and combine these as a mosaic to capture the entire object. (Image taken by Panahou School with Faulkes Telescope North operated by Las Cumbres Observatory Global Telescope Network. Used with permission.)

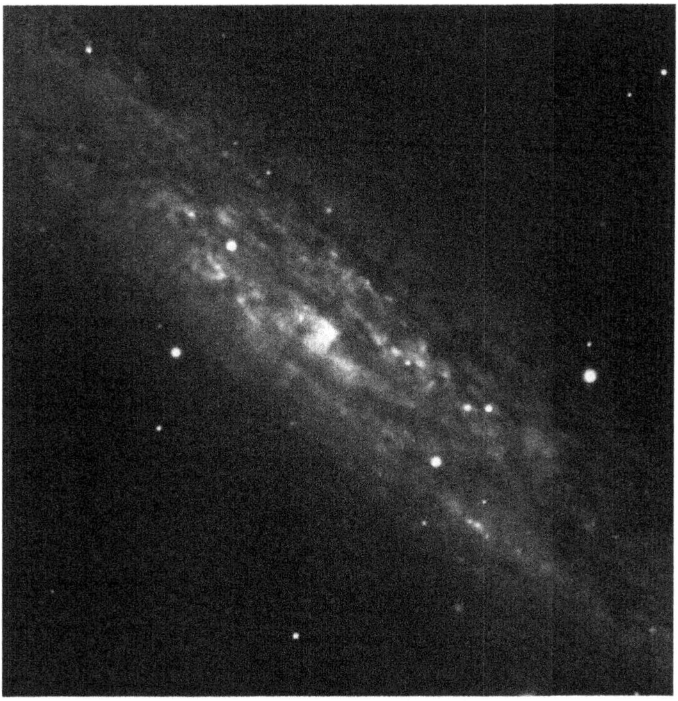

Fig. 8.3 NGC 253

NGC 281 Cassiopeia

RA: 00h 52m
Dec: +56.37
Size: 35′
Magnitude: 12.8
Filters: Hα, NB, RGB

This object is a fantastic HII region and can be observed visually in a 20-cm telescope as a faint ghostly glow against the stars, but shows up to great effect when photographed. Informally called the "Pacman Nebula" for obvious reasons once you image it, this large nebula includes the star cluster IC 1590, which may be captured in short exposures and then overlaid in *Photoshop* during the reduction process. Within NGC 281 are several dark clouds, or "Bok globules," in which new stars are being formed. Several fingers of dust intrude into the nebula from the periphery, but the most obvious feature is the dark lane of dust across the center. (Image taken by Martin Griffiths.)

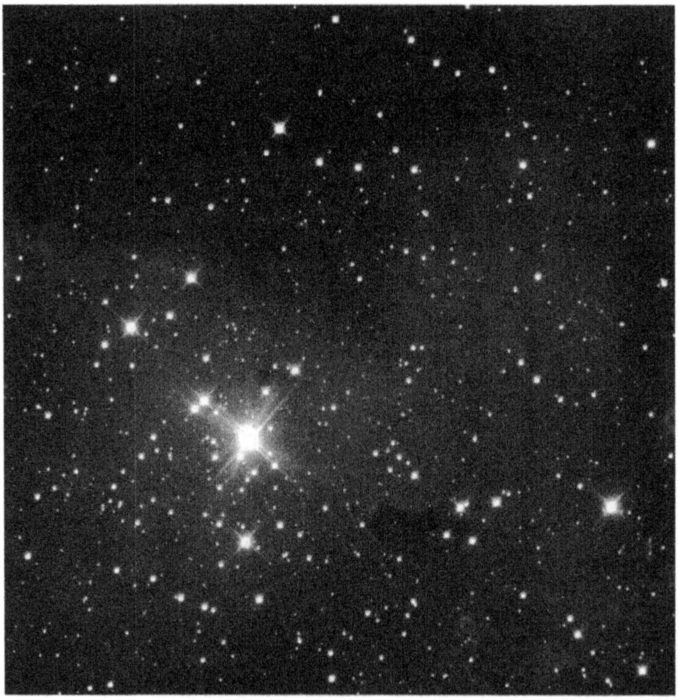

Fig. 8.4 NGC 281

Messier 33 Triangulum

RA: 01h 33m
Dec: +30.39
Size: 70×41′
Magnitude: 5.7
Filters: Hα, RGB

The incredible "Pinwheel Galaxy" is a favorite target of astrophotographers, as the object is richly detailed and has the added interest of containing some huge HII regions, one of which, NGC 604, is among the largest such nebulae ever found and makes a compelling object in its own right if the observer has a large telescope to separate it out from the body of M33. At a distance of 3.3 million light years this galaxy fills the field of most telescopes and will probably require a mosaic to be taken to get in the whole object. Although a little fainter than M31, it is still a bright object and easy to overexpose the central portion. (Image taken by Berlin International School with Faulkes Telescope North operated by Las Cumbres Observatory Global Telescope Network. Used with permission.)

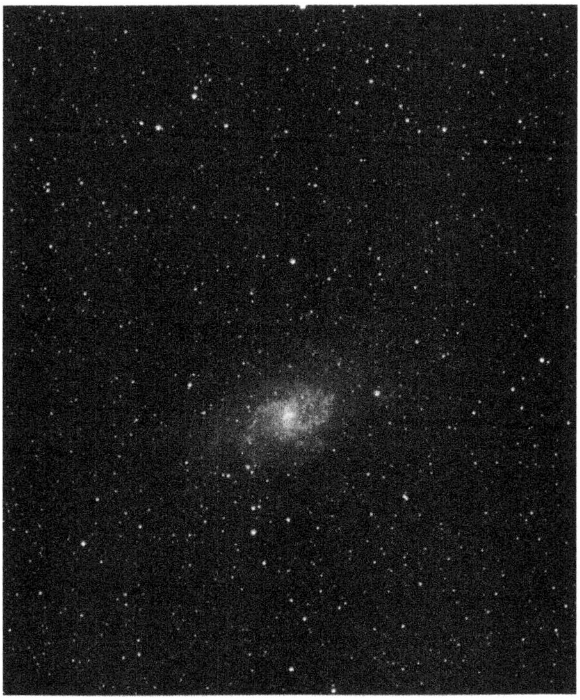

Fig. 8.5 M33

Messier 74 Pisces

RA: 01h 36m
Dec: +15.47
Size: 10×9′
Magnitude: 10.2
Filters: Hα, NB, RGB

At a distance of 32 million light years away, this is a fainter object than M33 above but is a little more concentrated and rewarding. It is a "grand design" galaxy, with clearly defined spiral arms and is face-on to us. This orientation, combined with its large size, gives M74 a relatively low surface brightness, though modern CCD cameras should have little problem in pulling out the required frames, and manipulation in *Photoshop* will bring out the best in this elusive visual object. This orientation also makes M74 suitable for an extragalactic supernova search, as most of its 100+ billion stars are available for scrutiny. (Image taken by Martin Griffiths with Faulkes Telescope North operated by Las Cumbres Observatory Global Telescope Network.)

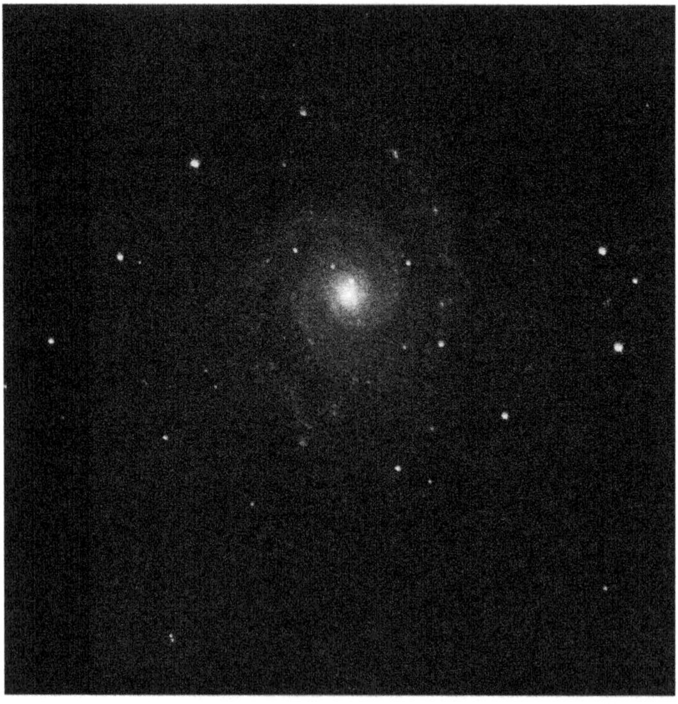

Fig. 8.6 M74

Messier 76 Perseus

RA: 01h 42m
Dec: +51.34
Size: 2.7′
Magnitude: 10.1
Filters: Hα, OIII, SII, RGB

This beautiful bipolar planetary nebulae was recorded in 1780 and has the epithet of the "Little Dumbbell Nebula." It was first thought to be a planetary nebula by Isaac Roberts in 1891, although Heber Curtis finalized this classification in 1918. Visually, M76 is elongated and evenly gray at first impression, resolving itself into a two-lobed structure oriented NW/SW with some faint markings on the northern lobe and the S one being slightly brighter. At higher magnifications the nebula is filled with light but no hint of color, although there is a distinct thicker edge to the S component. Imaging M76 presents few problems, as the nebula is bright and can be captured in a few seconds. Hα and OIII bring out some of the subtle structure surrounding the body, and RGB renders it in its appropriate colors. M76 is 2,500 light years away with a real diameter of 1.2 light years in extent. (Image taken by Martin Griffiths with Faulkes Telescope North operated by Las Cumbres Observatory Global Telescope Network.)

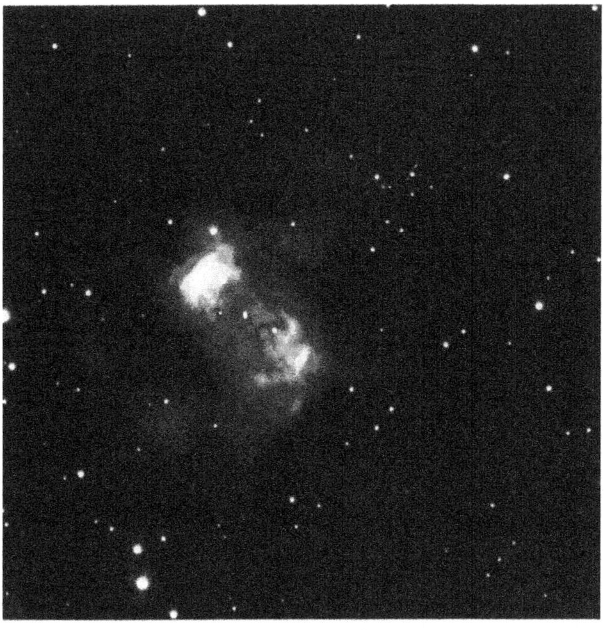

Fig. 8.7 M76

NGC 891 Andromeda

RA: 02h 22m
Dec: +42.20
Size: 13×2.5′
Magnitude: 10.8
Filters: Hα, RGB

A wonderful edge-on spiral galaxy, NGC 891 is a sliver of white light visually but reveals itself as an incredible splinter of golden light crossed by a delicate dust lane in long exposure photographs. At a distance of 27 million light years away it is a relatively easy target for RGB filtered photography, while Hα will bring out some of the details in the HII regions scattered above the dust lane, especially in the northern part of the galaxy. Recent investigation has shown this lovely galaxy to be a barred spiral, though this is not obvious due to the orientation of this object. (Image taken by Westminster School with Faulkes Telescope North operated by Las Cumbres Observatory Global Telescope Network. Used with permission.)

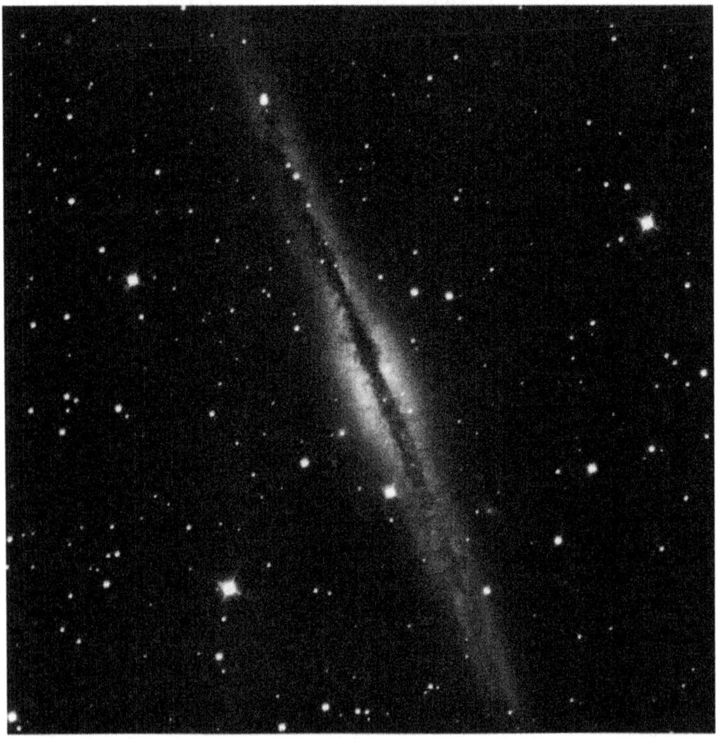

Fig. 8.8 NGC 891

IC 1805 Cassiopeia

RA: 02h 32m
Dec: +61.29
Size: 60×60′
Magnitude: 18.3
Filters: Hα, Hβ, RGB

The famous "Heart Nebula," enclosing the star cluster Melotte 15, is a favorite target for many observers but is a quite difficult object to image, as it has a very low surface brightness and is spread over a very large area of sky. A short focus telescope should capture the entire field around the cluster of two dozen or so stars and exposures in Hα and Hβ may bring out the best in this faint but rewarding object. IC 1805 is over 7,500 light years away on the Perseus arm of the Milky Way. In most exposures of 40–60 s the brightest part of the nebulae, NGC 896, will be readily visible as a round patch of light breaking into wave-like clouds of material. (Image taken by Martin Griffiths.)

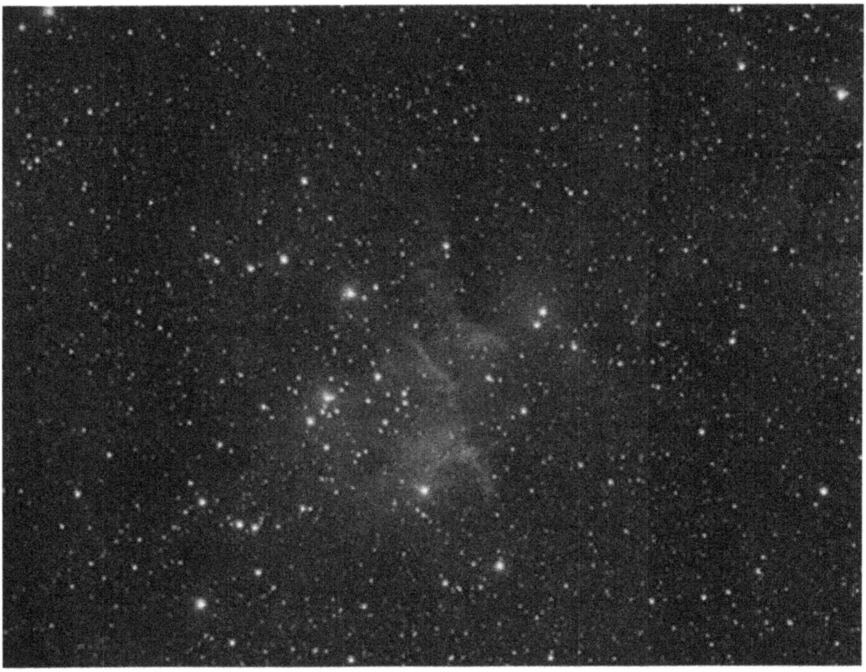

Fig. 8.9 IC 1805

NGC 1300 Eridanus

RA: 03h 19m
Dec: −19.24
Size: 6×4′
Magnitude: 11.4
Filters: Hα, NB, RGB

A fantastic barred spiral galaxy lying at a distance of 61 million light years, NGC 1300 is the stereotypical example of its class, and its face-on orientation and large size make it an ideal target to be imaged. It has a rich golden central nucleus and bar and two widely spaced arms that are speckled with bright luminous O-type stars, making a pleasing contract with the central portions. An Hα or even an NB filter may bring out the myriad of HII regions along the arms and dust lanes that extend right into the nucleus and will be noticed with careful manipulation of the image in *Photoshop*. (Image taken by Roger Manwood Grammar School with Faulkes Telescope South operated by Las Cumbres Observatory Global Telescope Network. Used with permission.)

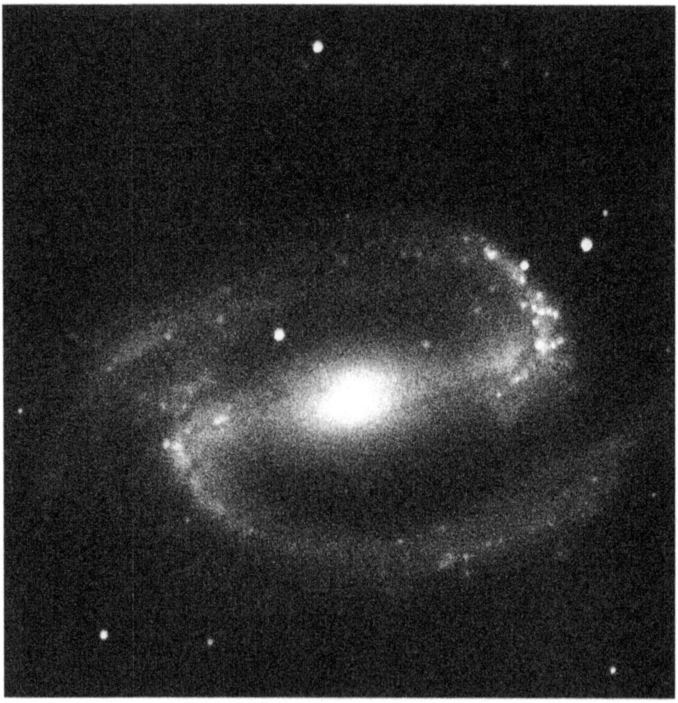

Fig. 8.10 NGC 1300

NGC 1333 Perseus

RA: 00h 47m
Dec: −11.52.3
Size: 6×3′
Magnitude: 10.8
Filters: NB, RGB

NGC 1333 is a small reflection nebula in Perseus, almost off the beaten track and seemingly far away from the main belt of the Milky Way. It is part of the Perseus molecular cloud and is one of the closest star-forming regions to Earth, with a distance of just over 1,000 light years. The cloud looks lumpy in the center with a filamentary tail and a dark bar of dust obscuring the southwestern portion of it. Careful imaging will bring out a few small areas of HII emission close to the dust bar. Although it will not be visible in the image, the star IRAS 4B here has a surrounding shell containing enough water vapor to fill the oceans of Earth five times over! Any forming planets in this region will make interesting research subjects with the presence of so much H_2O. (Image taken by Martin Griffiths with Faulkes Telescope North operated by Las Cumbres Observatory Global Telescope Network.)

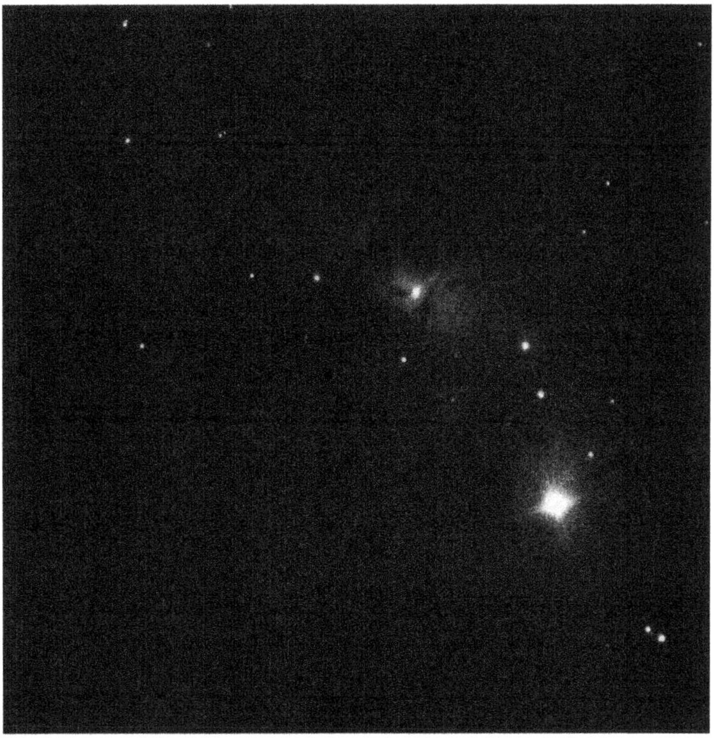

Fig. 8.11 NGC 1333

NGC 1360 Fornax

RA: 03h 33m
Dec: −25.52.3
Size: 8′
Magnitude: 9.4
Filters: Hα, OIII, SII, RGB

This is a faint, large nebula in Fornax that is not too difficult to find, but the planetary has a very low surface brightness, making it a little difficult to see with direct vision and is best imaged. Doing so brings out more detail, but there is little more than a smooth shell visible close to the ninth mag star. It is also rather low in the sky for observers in high northern latitudes, though those in the southern hemisphere will have no problems at all. The nebula responds moderately to use of an OIII filter, is elongated in a NE–SW direction and the sheer size of this planetary will render the object fainter than its quoted magnitude here. Imaging with OIII and Hα will bring out the best in this planetary. (Image taken by Sacred Heart Catholic School with Faulkes Telescope South operated by Las Cumbres Observatory Global Telescope Network. Used with permission.)

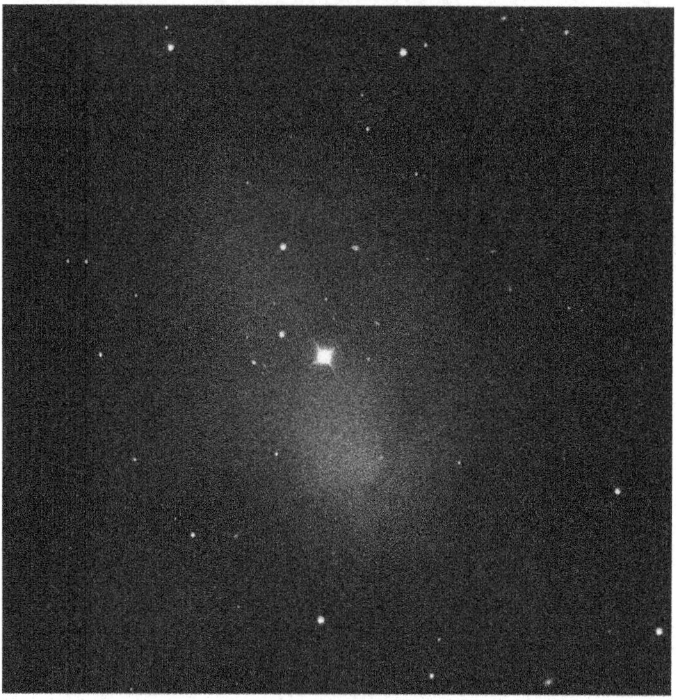

Fig. 8.12 NGC 1360

Messier 45 Taurus

RA: 03h 47m
Dec: 24.07
Size: 110′
Magnitude: 3.0
Filter: RGB

The superb Pleiades star cluster is a regular target for astronomers and is one of the most obvious astronomical objects in the sky, remarked upon by every single culture in human history. In addition to the plethora of stars, there is the subtle filigree of dust clouds and a reflection nebula that surrounds the group, and these are at their best near Merope and Maia, where the electric blue reflection from the main group of stars blazes off the filaments of dust surrounding the cluster. RGB filters or UBVRI will bring out the best of this group, which should be captured with a short-focus telescope to get the whole cluster in the field. Exposures will be quite short, as the cluster and nebula are quite bright. (Image taken by Martin Griffiths.)

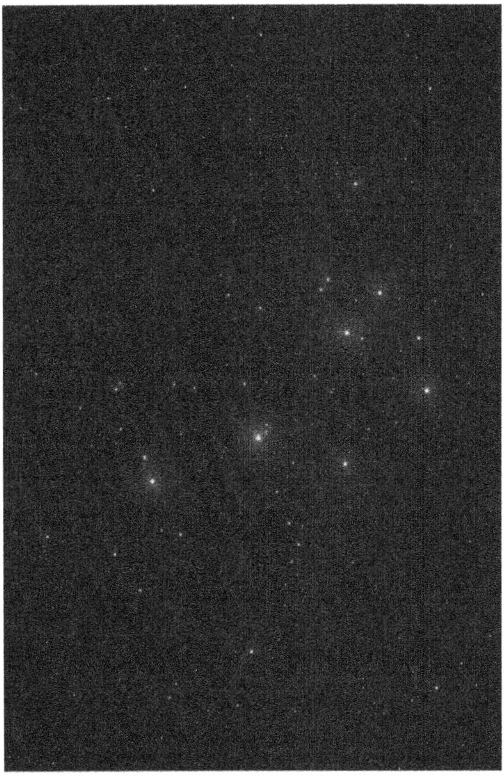

Fig. 8.13 M45

NGC 1491 Perseus

RA: 04h 03m
Dec: +51.40
Size: 10×5′
Magnitude: 8.6
Filters: Hα, RGB

This emission nebulae is a small cloud in the northern section of Perseus close to a seventh magnitude star and shows up in short exposures as a veil-like smear of light close to the star, but longer exposures bring out a rounded extension about 10′ in extent. Lying almost 10,000 light years away, NGC 1491 is bright enough to be seen in a moderate telescope, but CCD imaging in Hα or with narrowband filters will bring out the best in this delicate object. Lying close by is the emission nebulae Sharpless 2-209, which should be visible in the large view of a telephoto lens or a short-focus telescope. (Image taken by Martin Griffiths with Faulkes Telescope North operated by Las Cumbres Observatory Global Telescope Network.)

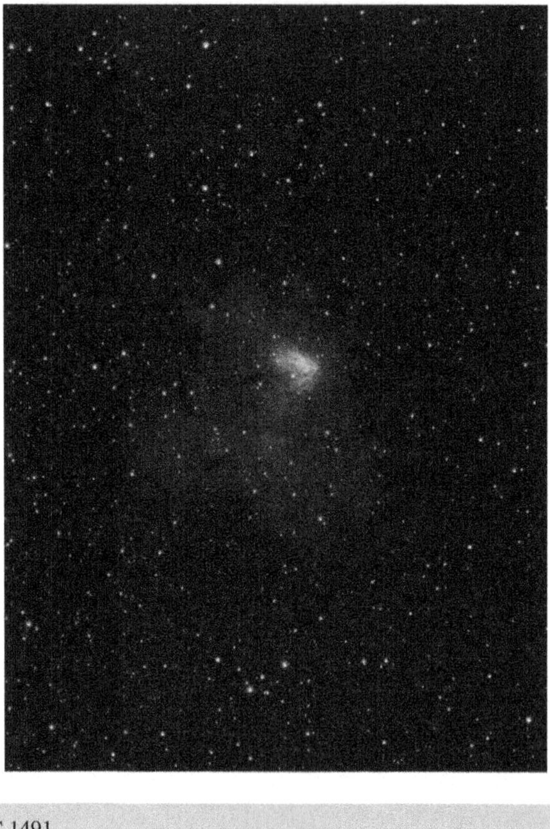

Fig. 8.14 NGC 1491

NGC 1499 Perseus

RA: 04h 03m
Dec: +36.25
Size: 160×40′
Magnitude: 6.4
Filters: Hα, Hβ, RGB

The wonderful complex of the "California Nebula" is difficult to see visually and requires some skill and patience to photograph. At almost 3° in extent this large, low surface brightness object is about 1,500 light years from Earth and is best seen in Hβ. The star ξ Persei is probably responsible for its illumination and is part of an O–B association and extended nebulosity in the region that also includes NGC 1333 to the SW. Many photographers have used DSLRs to capture this object due to its very large size. (Image taken by Martin Griffiths.)

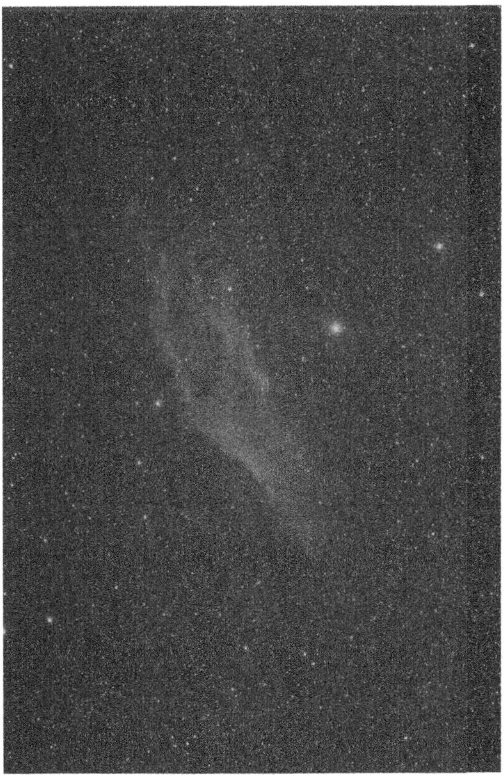

Fig. 8.15 NGC 1499

NGC 1501 Camelopardalis

RA: 04h 06m
Dec: +60.55.2
Size: 56″
Magnitude: 11.4
Filters: Hα, OIII, SII, RGB

This is a beautiful planetary nebula that is a visual "deep-sky double" in a low power field, with the star cluster NGC 1502 at the end of "Kemble's Cascade." It's also a Herschel 400 object, which reveals itself as a smooth blue-gray, round, well defined object that seems even across its surface with no central star visible. For the photographer, the nebula responds well to NB and OIII use and is a mottled ovoid of green-gray light with a central knot that hints at the white dwarf at the center. (Image taken by Chantal Barboram with Faulkes Telescope North operated by Las Cumbres Observatory Global Telescope Network. Used with permission.)

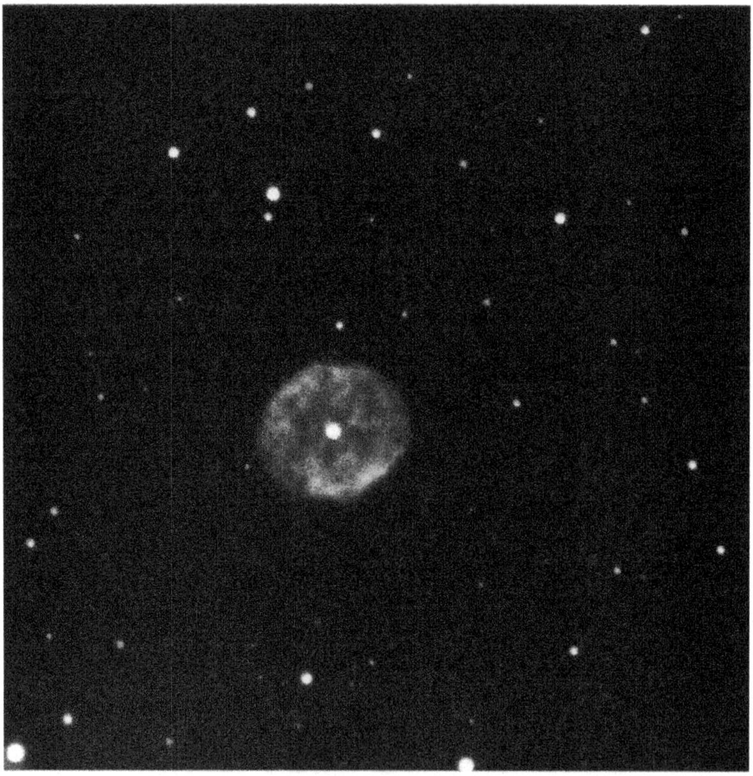

Fig. 8.16 NGC 1501

NGC 1514 Taurus

RA: 04h 09m
Dec: +30.46.6
Size: 130″
Magnitude: 10.9
Filters: Hα, OIII, SII, RGB

This is known as the "Crystal Ball Nebula" and is a rather difficult object due principally to the bright central star, which drowns out the nebula. It's also in a relatively barren location in the sky, though NGC 1514 reveals itself as an amorphous concentric gray wash of milky gas that CCD imaging brings out fairly well, though a little burn out and overexposure of the central portion is inevitable. It was this object that enabled William Herschel to link nebulae and stars together, paving the way for future developments in many fields of astrophysics. (Image taken by Ysgol Uchradd Tywyn with Faulkes Telescope North operated by Las Cumbres Observatory Global Telescope Network. Used with permission.)

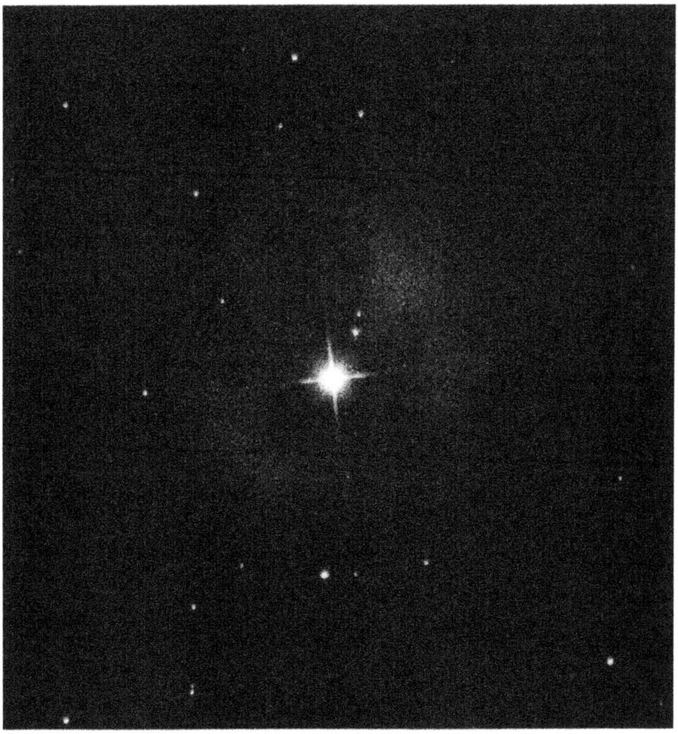

Fig. 8.17 NGC 1514

NGC 1535 Eridanus

RA: 04h 14m
Dec: −12.44.4
Size: 50″
Magnitude: 9.6
Filters: Hα, OIII, SII, RGB

"Cleopatra's Eye" is a fabulous planetary in Eridanus and a real showpiece winter object. The nebula is a smooth blue-green shell of light over 1′ across, a little elliptical with the axis lying NE/SW but no central condensation and no hint of a central star. It shows great structure if the correct filters and short exposures are used, and the images can be manipulated in *Photoshop* with some ease. NGC 1535 is also one of the Herschel 400 objects and is one of the best planetary nebula targets of the NGC catalog for any small telescope. (Image taken by Simon Langton Grammar School with Faulkes Telescope North operated by Las Cumbres Observatory Global Telescope Network. Used with permission.)

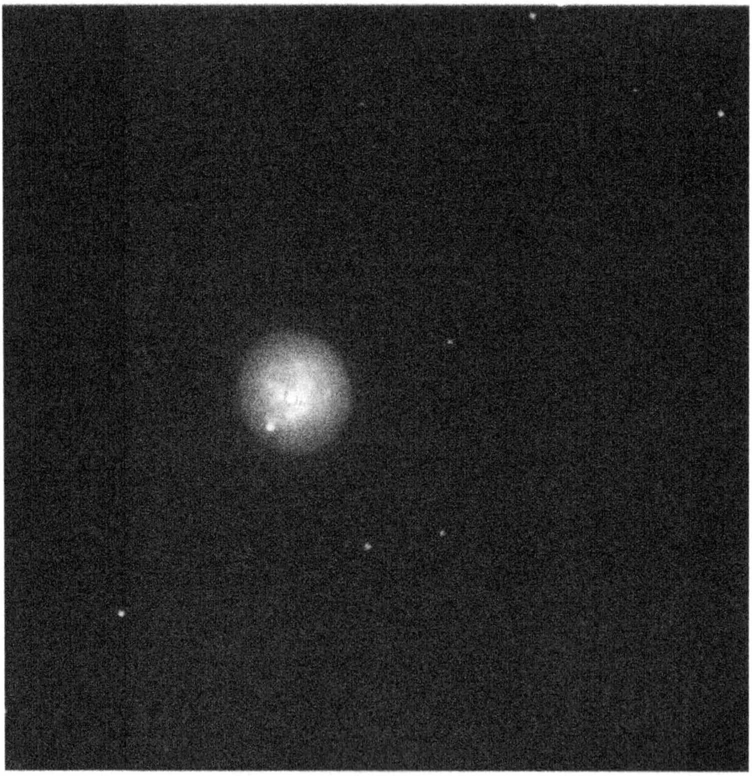

Fig. 8.18 NGC 1535

NGC 1554 Taurus

RA: 04h 21m
Dec: +19.32
Size: 5′
Magnitude: 12.3
Filters: NB, RGB

This small emission nebula in Taurus is a challenging object, as the cloud of gas surrounding the star T Tauri is small and closely connected with the bright luminary. It was first discovered in the nineteenth century by the observer J.R. Hind and has been known ever since as "Hind's Variable Nebula." It is an interesting object in that it is the birth shroud of a sun-like star and is quite active, sending streamers of gas and dust millions of kilometers into space. It is a bright infrared object, but for purposes of this book both NB and RGB filters should pick it up, though be careful to avoid the main star overexposing.

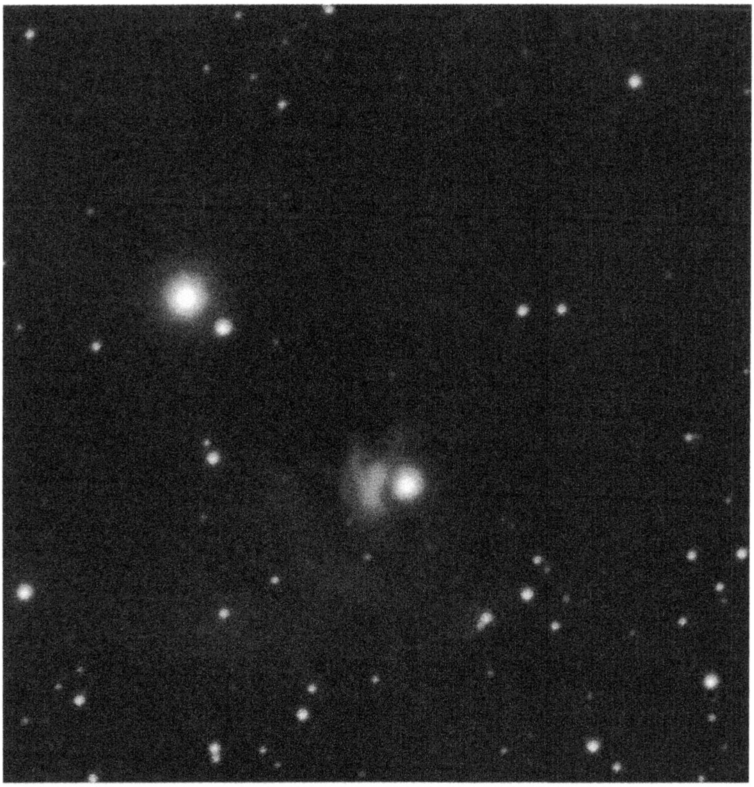

Fig. 8.19 NGC 1554

IC 2118 Eridanus

RA: 05h 02m
Dec: −07.54
Size: 180×60′
Magnitude: 13.1
Filters: NB, RGB

The fabulous "Witch Head Nebula" is a dim reflection nebula possibly illuminated by the bright star Rigel over the constellation border in Orion. Shining with a blue hue and being large in size, the Witch Head is quite dim and requires a set of long exposures to bring out the best in the object. IC 2118 lies around 900 light years away and is the result of the compression of the interstellar medium by the high-speed stellar winds from the Orion O–B association. A zoom lens on a driven DSLR or a short-focus telescope should give a wide enough field to capture this elusive cloud. (Image courtesy of Wikipedia: http://en.wikipedia.org/wiki/File:Reflection.nebula.arp.750pix.jpg.)

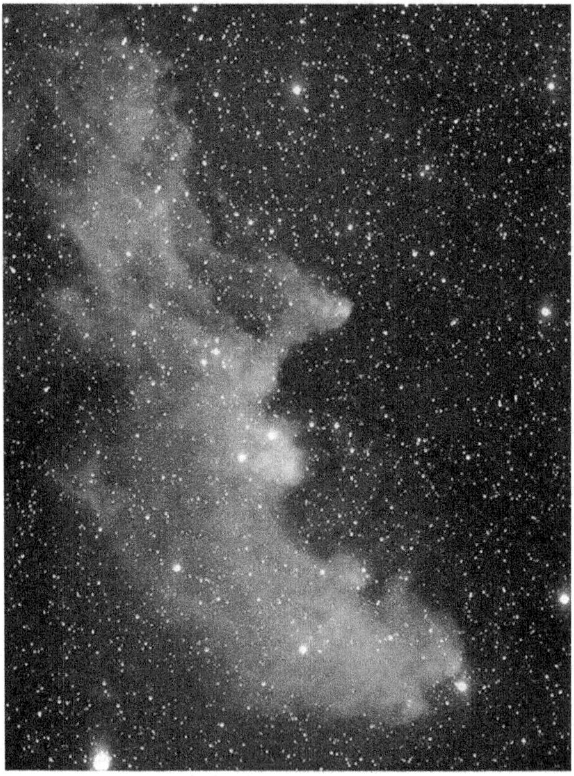

Fig. 8.20 IC 2118

NGC 1788 Orion

RA: 05h 06m
Dec: −03.21
Size: 8 × 8′
Magnitude: 9.2
Filters: Hα, NB, RGB

This is a small but often neglected bright reflection nebula in the southwestern corner of Orion and is a target for small telescopes come the winter months but is overlooked in the rush to get to its bigger brother M42/43. NGC 1788 is blockish, almost square in appearance, with a number of bright stars superimposed across its feathery tendrils of light yet revealing some lovely detail of contrast, star colors and dust lanes in combined images. Stars fade rapidly on its southern perimeter as dark unlit dust clouds populate this region. (Image taken by Martin Griffiths.)

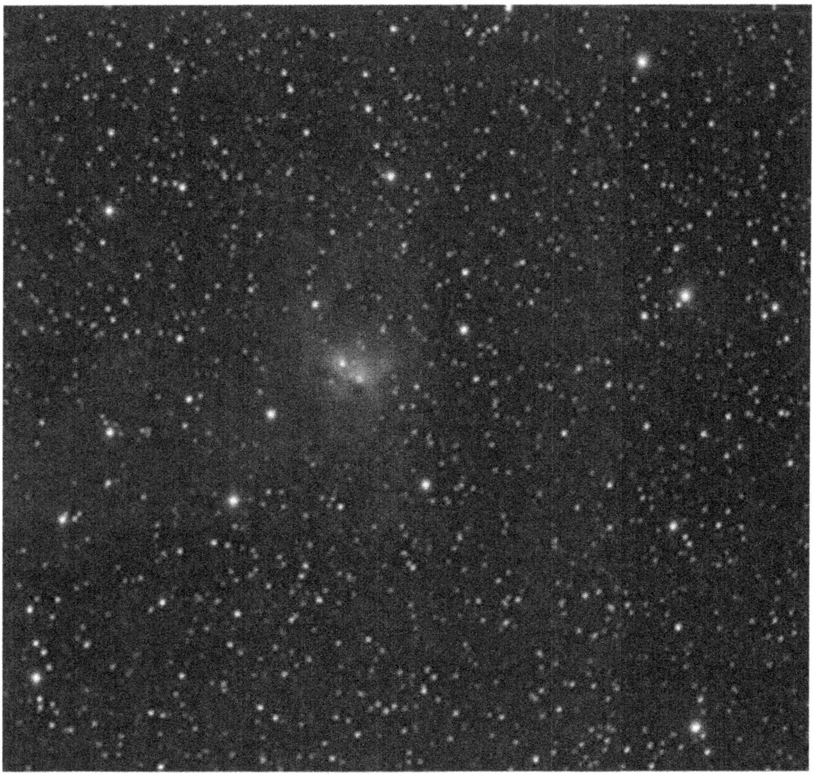

Fig. 8.21 NGC 1788

IC 410 Auriga

RA: 05h 22m
Dec: +33.23
Size: 60×60′
Magnitude: 11.1
Filters: Hα, NB, RGB

These two large HII regions are superb targets for Hα photography. Both nebulae are relatively faint and very difficult to observe visually, but their proximity to the star clusters M38 and NGC 1907 give the photographer a clear target for a wide-field shot with either a short-focus telescope or telephoto lens. IC 410 lies to the SE of a trail of stars 14, 16 and 19 Aurigae, which are naked-eye objects, while IC 405 lies to their NW and is the larger of the two clouds of gas. Although they are common astrophotography targets, their faintness and extent make them a good challenge. The Milky Way in this region is wonderfully bright but is clearly dimmed by the presence of these gas clouds and their associated dust. (Image taken by Martin Griffiths.)

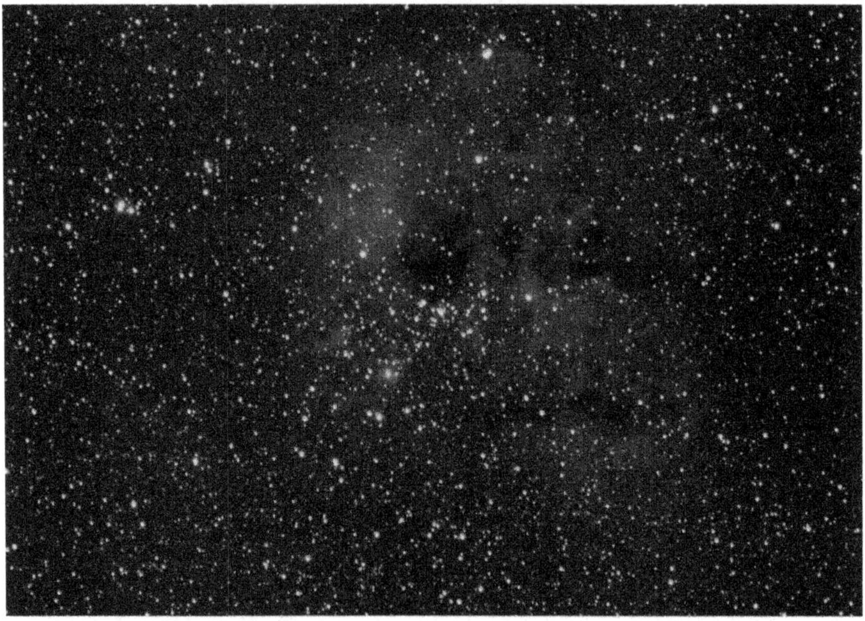

Fig. 8.22 IC 410

NGC 1931 Auriga

RA: 05h 31m
Dec: +34.15
Size: 3 × 3′
Magnitude: 10.1
Filters: Hα, NB, RGB

This lovely little nebula to the SE of the star cluster M37 is a small roundish object shining in the light of Hα and reflected starlight. At 7,000 light years away this tiny nebula is clearly quite large in real terms and is lit by a small cluster of four stars, which resemble the Trapezium in M42 but on a much dimmer and smaller scale. These stars are generally burned out on most images, as the fainter parts of the nebula become the preferred focus of the image. (Image taken by Oliver Tunnah with Faulkes Telescope North operated by Las Cumbres Observatory Global Telescope Network. Used with permission.)

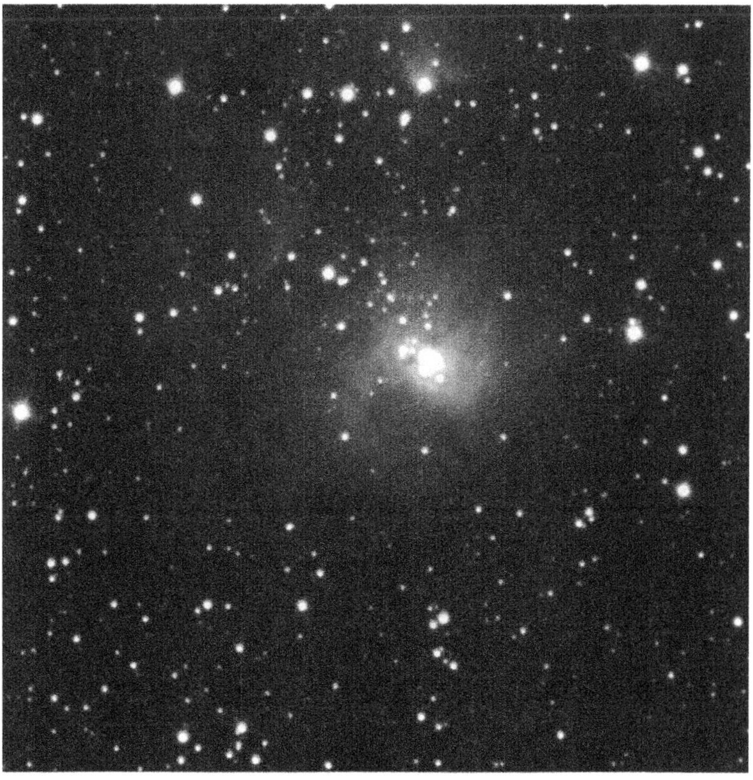

Fig. 8.23 NGC 1931

Messier 1 Taurus

RA: 05h 34m
Dec: +22.00
Size: 7×5′
Magnitude: 8.4
Filters: Hα, NB, RGB

The superb "Crab Nebula" is a supernova remnant in Taurus and a firm favorite of photographers and observers alike. The star responsible for the cloud exploded over 6,000 years ago, but its huge distance rendered the object visible only a millennium ago. It is relatively easy to image in visible light, but a nebula filter coupled with Hα may bring out the fainter extensions of the crab's claws and the subtle detail that crosses the interior of this wonderful object. The light from the synchrotron radiation filling the shell is blue-white in color and contrasts beautifully with the yellow and red filaments of the outer shells of gas. As a challenge, see if you can image the bright central pulsar. (Image taken by Rosie Cane with Faulkes Telescope North operated by Las Cumbres Observatory Global Telescope Network. Used with permission.)

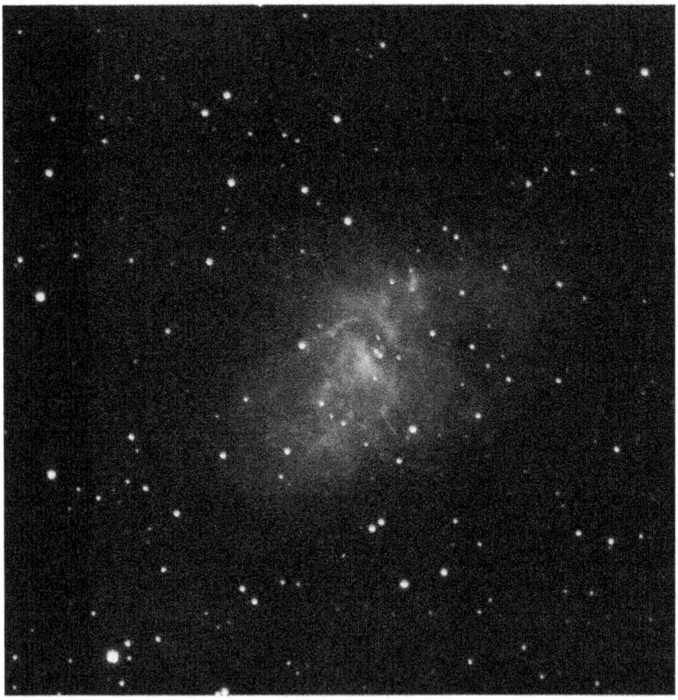

Fig. 8.24 M1

Messier 42/43 Orion

RA: 05h 35m
Dec: –05.23
Size: 70×60′
Magnitude: 4.1
Filters: Hα, NB, RGB

The incredibly detailed and large Orion Nebula complex is a wonderful target, and due to its high surface brightness presents few problems, apart from burn out of the central area around the Trapezium. There is a wealth of detail in the outer regions of the cloud and a sufficiently short-focus telescope or telephoto lens will also capture NGC 1977 to the north. The range of color and the detail that can be acquired in the image is incredible and leads to many hours of manipulation in *Photoshop* to bring out everything visible in this nebula. The dust cloud separating M43, or "De Mairan's Nebula," from M42, the main body of the cloud, is clear and reveals lovely detail with careful processing. The entire complex is about 1,300 light years away and is a bubble of material. (Image taken by Martin Griffiths.)

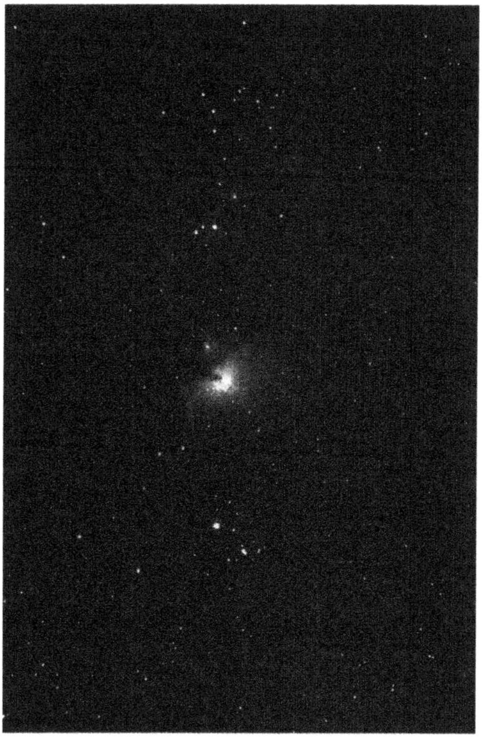

Fig. 8.25 M42/43

NGC 1977 Orion

RA: 05h 35m
Dec: −04.48
Size: 40×25′
Magnitude: 7.1
Filters: Hα, NB, RGB

Among a fantastic array of nebulae and faint stars to the north of the main Orion Nebula complex in the Hunter's belt is the "Running Man Nebula," a wispy yet bright object that is a common target for digital SLRs and CCD cameras. The whole cloud has several NGC designations, as it appears to be broken by occasional dust clouds. Short exposures in Hα should bring out the figure of the running man, and RGB/UBVRI will allow the details of this sometimes neglected nebula to come out well. Not as overwhelming as its bright neighbor to the south, NGC 1977 is a great area of contrast and lies at a distance of 1,300 light years with a spread of over 30 light years in area at this distance. It is part of the huge Orion molecular cloud complex, generating many new stars in this region. (Image taken by Martin Griffiths.)

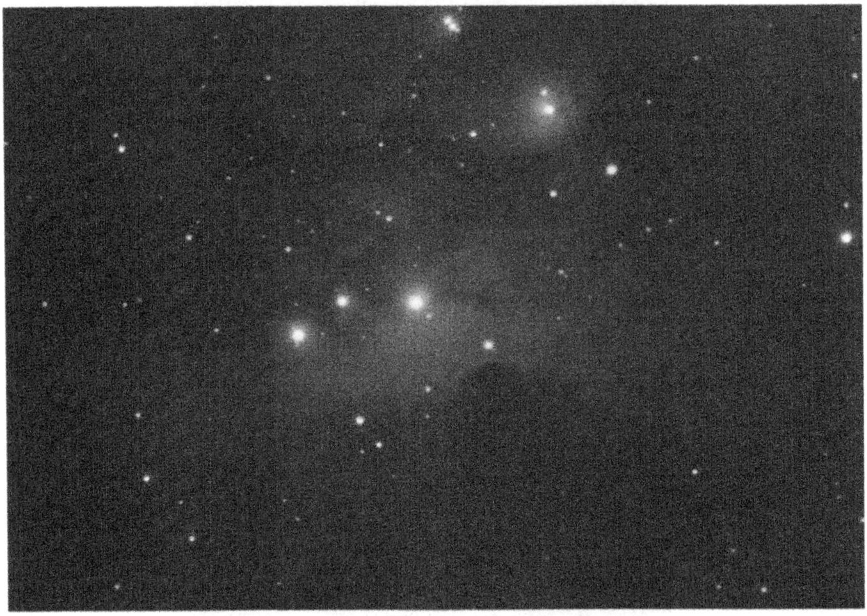

Fig. 8.26 NGC 1977

Simeis 147 Taurus

RA: 05h 41m
Dec: +27.46
Size: 210×200′
Magnitude: 12.5
Filters: Hα, NB, RGB

This is revealed as a wonderfully intricate supernova remnant in large telescopes and is a challenge for amateur observers simply due to its large size. At over 3° in diameter, Simeis 147 is difficult to cram into the field of a short-focus telescope and may require the services of a telephoto lens instead. A filamentary cloud known colloquially as the "Spaghetti Nebula" is uncovered by careful photography. Amazingly, the whole complex is not too far away from another great SN remnant, the Crab Nebula, M1. Simeis 147 is about 3,000 light years away, and from the size of the cloud, it is estimated that this star blew up 40,000 years ago. (Image taken by Martin Griffiths.)

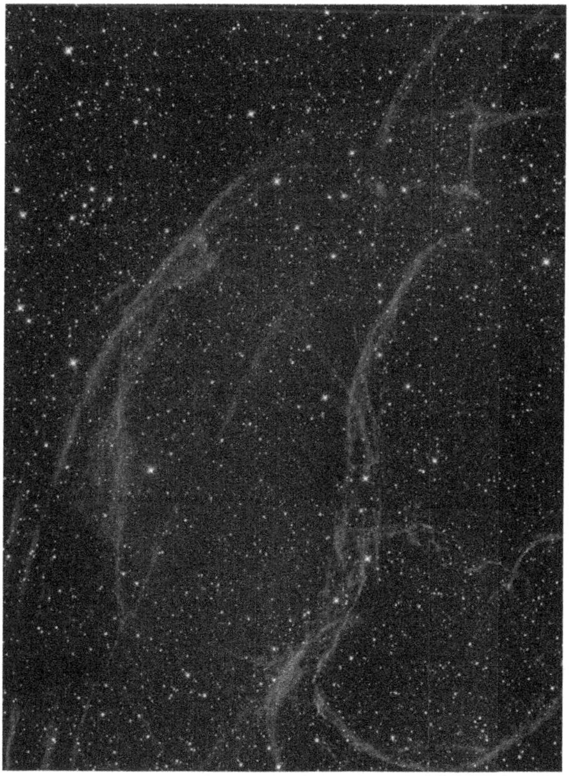

Fig. 8.27 Simeis 147

NGC 2024 Orion

RA: 05h 41m
Dec: −01.51
Size: 30×30′
Magnitude: 5.6
Filters: Hα, NB, RGB

The "Tank Track," "Flame" or "Flaming Heart Nebula" is another part of the Orion molecular cloud complex and is easy to find lying just to the east of the main stars of Orion's Belt and is possibly illuminated by Alnitak, the first eastern belt star. The nebula is fairly bright and responds well to photography with filters, which then bring out a heart-shaped cloud crossed by a lovely contrasting lane of dust with indentations, making the whole thing look like it's crossed by a tank track. In a short-focus telescope or telephoto lens, the whole object merges into our next target, IC 434, but is still distinct as a pall of dust separating the two clouds, which should stand out on your image. Like much of the Orion cloud complex, the distance to NGC 2024 is about 1,300–1,500 light years away. (Image taken by Martin Griffiths.)

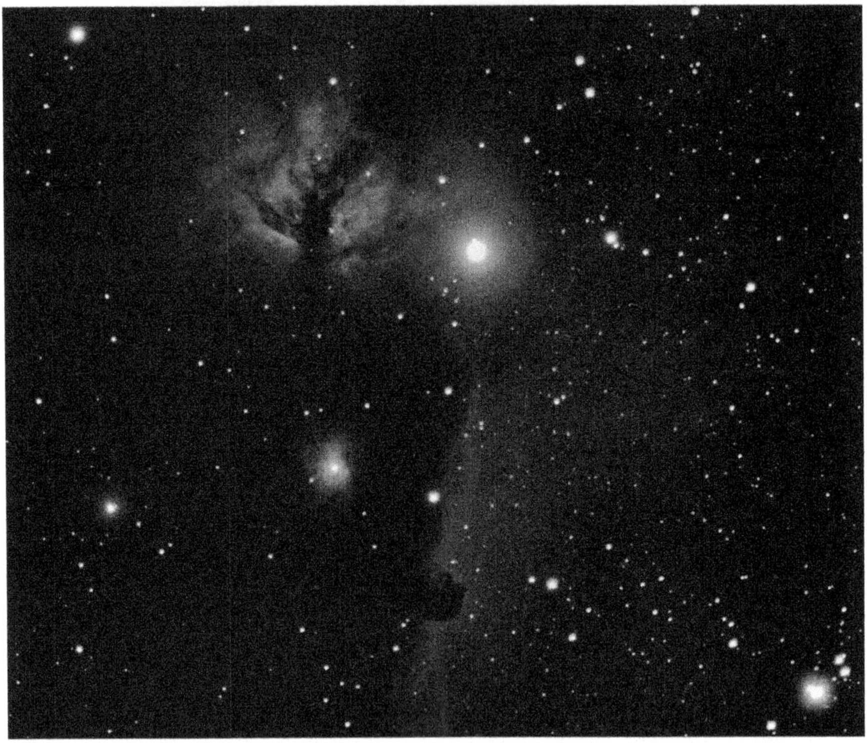

Fig. 8.28 NGC 2024

IC 434 Orion

RA: 05h 41m
Dec: −02.03
Size: 40 × 20′
Magnitude: 7.2
Filters: Hα, NB, RGB

This faint red cloud of hydrogen gas extends southward from Alnitak in Orion's Belt and contains one of the most sought-after photography targets, B33, or the "Horsehead Nebula." This dark band of dust in the shape of a chess piece knight is actually over a light year in extent and is backlit by the ghostly glow of IC 434, which soon dissipates into the dust of the molecular clouds in this region. The materials of the cloud contain complex organic molecules and was recently been found to contain amino acids, the base building blocks of life, too. At 1,500 light years away, IC 434 is a beautiful object for study and the use of Hα and nebular filters should bring out the faint details in this amazing area. (Image taken by Martin Griffiths.)

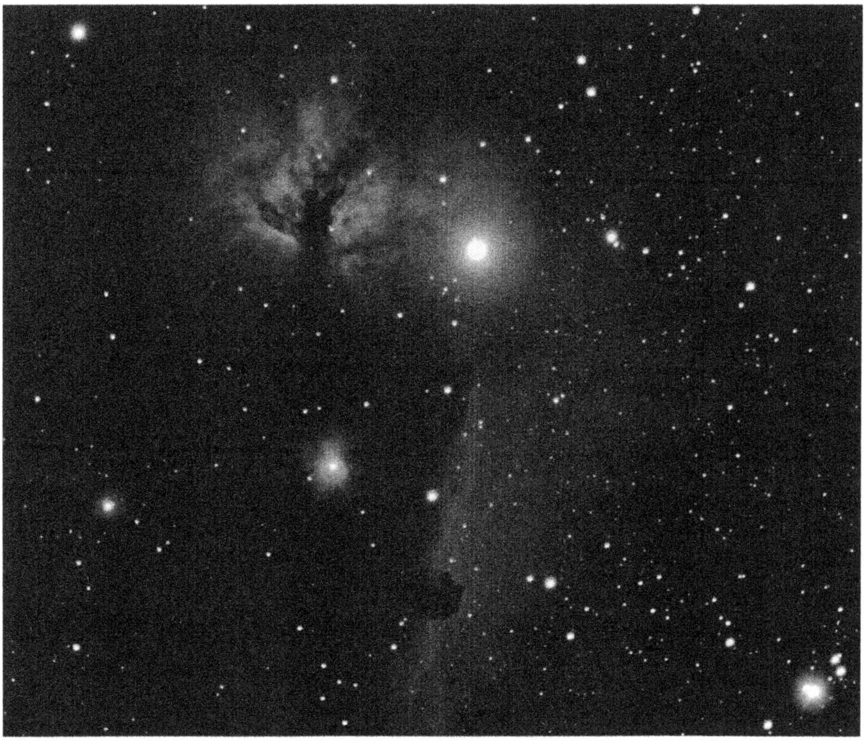

Fig. 8.29 IC 434

Messier 78 Orion

RA: 05h 46m
Dec: +00.03
Size: 8×6′
Magnitude: 78.3
Filters: Hα, NB, RGB

This is a small cloud to the north of Orion's Belt and is a tempting target but needs careful photography, as many of the details are faint, and the two seemingly disconnected clouds of this reflection nebula are in fact well connected and have a great dust lane crossing the field, too. M78 contains two tenth magnitude stars, which are responsible for the illumination of the blue reflection nebula. In all, 45 T Tauri-type stars have been discovered here in addition to a plethora of Herbig-Haro infrared objects beyond the scope of most ground-based observatories. Again, as part of the Orion molecular cloud complex, the distance to M78 is around 1,500 light years, and the cloud extends to 5 light years at this distance. (Image taken by Victoria College with Faulkes Telescope North operated by Las Cumbres Observatory Global Telescope Network. Used with permission.)

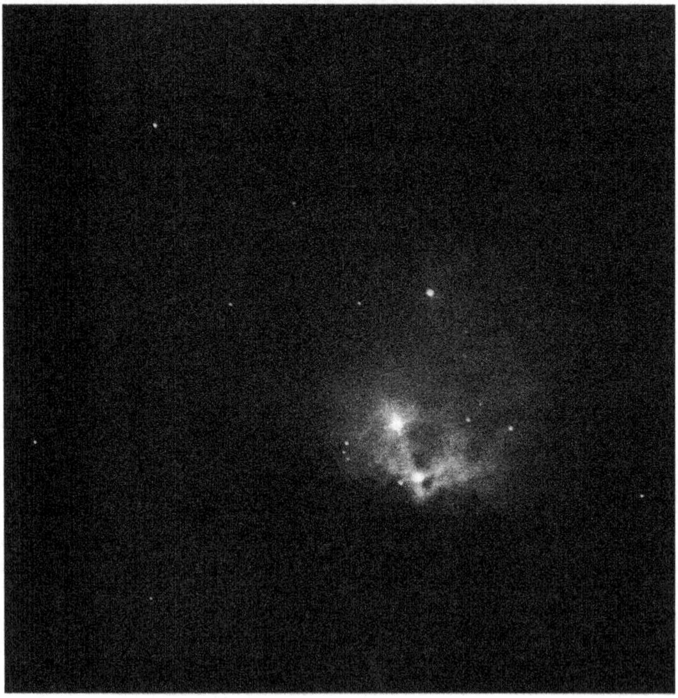

Fig. 8.30 M78

NGC 2070 Doradus

RA: 05h 38m
Dec: −69.05
Size: 40×25′
Magnitude: 7.6
Filters: Hα, NB, RGB

The incredible "Tarantula Nebula" in the Large Magellanic Cloud. The whole area is a knot of light in binoculars, but a camera and telescope bring out the true beauty of this awesome gas cloud. Lying at 160,000 light years away the Tarantula Nebula is so luminous that it would cast shadows if it were at the same distance as the Orion Nebula! In the heart of the nebula is the star cluster R136 that was, at one time (up to the advent of the Hubble Space Telescope), thought to be the single star 30 Doradus. Northern hemisphere viewers will chomp at the bit at not being able to see this fantastic object, but southern observers can make the most of this wonderful object that will fit well into the fields of most telescopes and CCD cameras. If not, a mosaic of the area will be a necessary but worthwhile exercise. (Image taken by Antoni Kaspryzk with Faulkes Telescope South operated by Las Cumbres Observatory Global Telescope Network. Used with permission.)

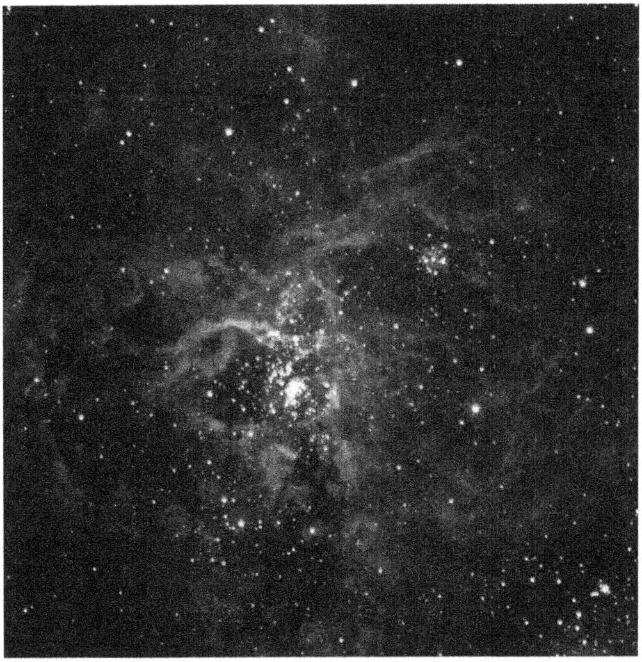

Fig. 8.31 NGC 2070

NGC 2170 Monoceros

RA: 06h 07m
Dec: −06.23
Size: 2×2′
Magnitude: 711.1
Filters: Hα, NB, RGB

This small reflection nebula in the constellation Monoceros was discovered on an autumn evening in 1784 by William Herschel. At a distance of 2,400 light years the nebula is quite small but can be well resolved and has a dust lane dividing it vertically with another lane approaching it on a horizontal plane. The center is quite messy and feathery, but it still makes a fair object to photograph and can be explored visually with a small reflector. It is part of a large molecular cloud named the Monoceros R 2 association and is part of a complex of nine other reflection nebulae in this area all illuminated by B-type stars of ages varying between 6 and 10 million years. The core of this cloud contains one of the most energetic bipolar outflows from a newly formed star, which is slowly destroying this small but lovely object.

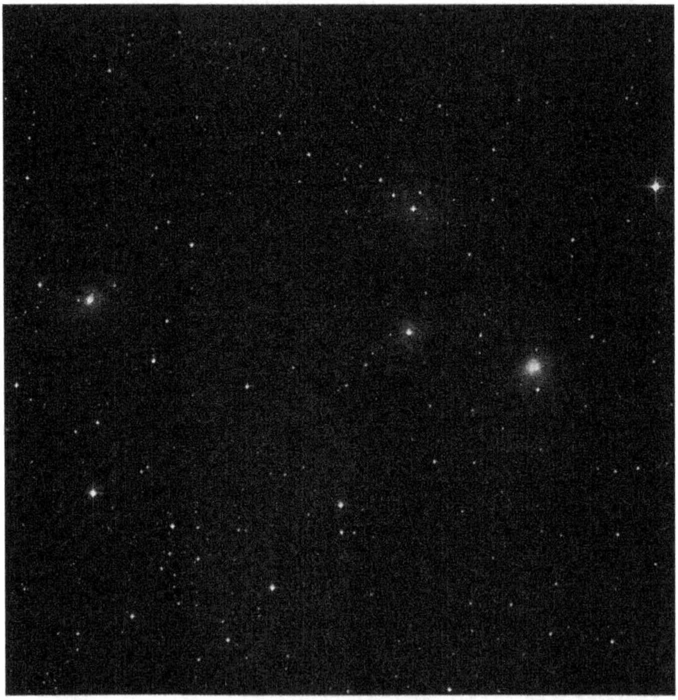

Fig. 8.32 NGC 2170

IC 443 Gemini

RA: 06h 17m
Dec: +22.31
Size: 20 × 15′
Magnitude: 11.9
Filters: Hα, NB, RGB

The so called "Jellyfish Nebula" in Gemini is actually a supernova remnant located about 5,000 light years from Earth and has a full size of 50′ once all the filamentary structure is accounted for. Unlike other SN remnants like the Crab Nebula, the cloud is shining due to thermal emission rather than being shocked and heated by a neutron star. There is such an object at the apex of the nebula, which is surrounded by a shell of material undergoing synchrotron emission, but this going to be very hard to spot from an amateur point of view. Nevertheless the "half-moon" shape of the nebula and its relative faintness make it a great object to try and capture.

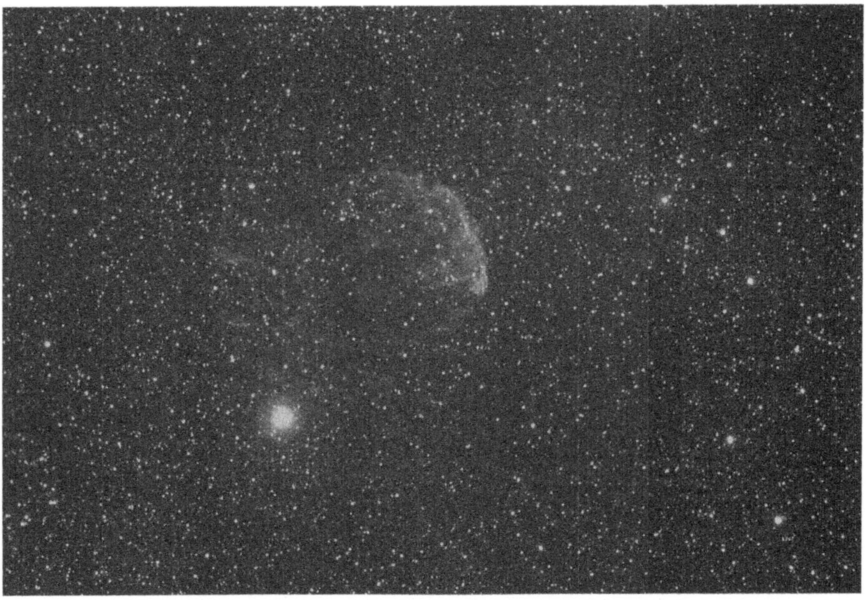

Fig. 8.33 IC 443

NGC 2237 Monoceros

RA: 06h 33m
Dec: +04.59
Size: 80 × 80′
Magnitude: 9.1
Filters: Hα, NB, RGB

This is a large complex of nebulae that have been subdivided into several NGC numbers: NGC 2238, 2239, 2246 and 2244. It is actually the star cluster at the center of the great "Rosette Nebula" in Monoceros. It is a huge object almost three times the diameter of the Moon and will probably require a mosaic of images unless the observer has a short-focus refractor and a wide field for his or her CCD. Lying over 5,000 light years away this large and spectacular complex draws the attention of astrophotographers every winter, and its proximity to the celestial equator makes it a target no matter what your latitude. The central cluster is almost a cuboid of stars, which reminds some of a box kite! These O and B stars are responsible for illuminating the nebula, which is riven with dark globules and small dust lanes that show up very well in even short exposures. (Image taken by Martin Griffiths.)

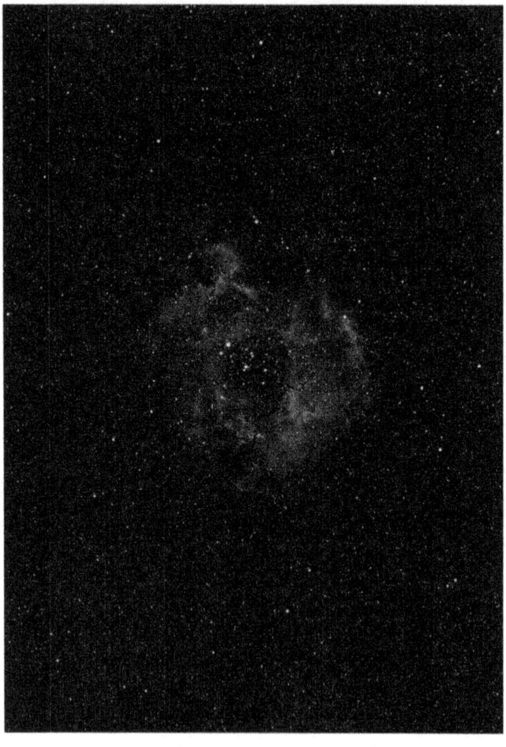

Fig. 8.34 NGC 2237

NGC 2261 Monoceros

RA: 06h 39m
Dec: +08.45
Size: 2×2′
Magnitude: 9.0
Filters: Hα, NB, RGB

The small comet shaped reflection nebula surrounding the star R Monocerotis was christened "Hubble's Variable Nebula," after he studied its light output at Yerkes and Mt. Wilson observatories. He chose this object for the "first light" of the Palomar 200-in. telescope in 1949. The cloud is illuminated dust obscuring a newly formed star and bears some similarities to NGC 1555, Hinds Variable Nebula in Taurus (T Tauri), as a similar process is occurring, with a newborn star shedding its formative shrouds of gas and dust—though in this case R Monocerotis is a much larger star than T Tauri. It is believed that the fan-shaped cloud has a counterpart opposite its axis that is obscured by thick dust in the region. Therefore this object could have been much larger if not for this obscuration. The nebula varies between magnitude 9.5 and 11 in an uneven pattern but is worth recording simply for its relative rarity. (Image taken by Paulet High School with Faulkes Telescope North operated by Las Cumbres Observatory Global Telescope Network. Used with permission.)

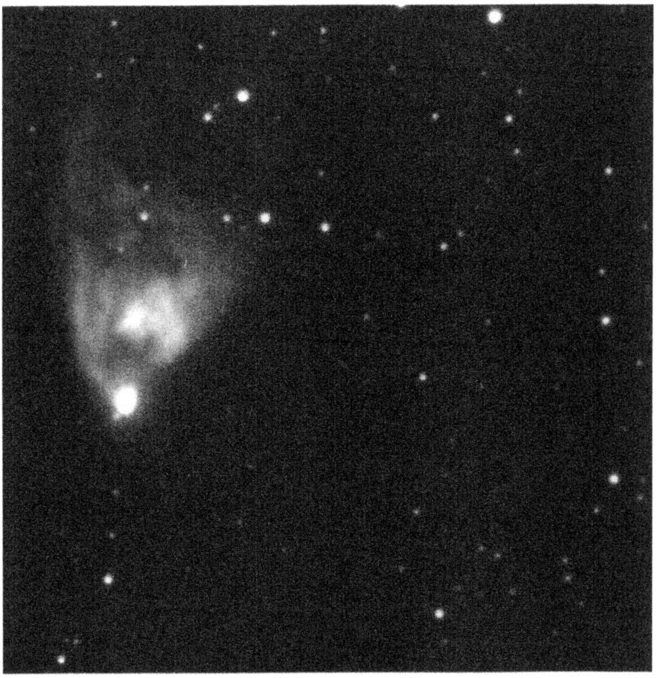

Fig. 8.35 NGC 2261

NGC 2264 Monoceros

RA: 06h 41m
Dec: +09.53
Size: 20×20′
Magnitude: 4.3
Filters: Hα, NB, RGB

The fantastic "Cone Nebula," which contains the star cluster commonly known as the "Christmas Tree," lies at a distance of 2,700 light years from Earth in the outer arms of our Milky Way Galaxy. It has been called the "Jesus Christ Nebula," since in long exposures it looks like a praying man with a halo of stars! As an HII region an Hα filter should bring out much of the complex details, and RGB will then render colors and smaller details visible, especially the dark shroud of dust that fans out into space before disappearing into the blackness of the night. The cluster itself has S Monocerotis as the "trunk," and the group then spreads out like an arrowhead in a glitter of starlight framing a shape that does look remarkably like the lights glowing on a traditional Christmas tree. (Image taken by Martin Griffiths.)

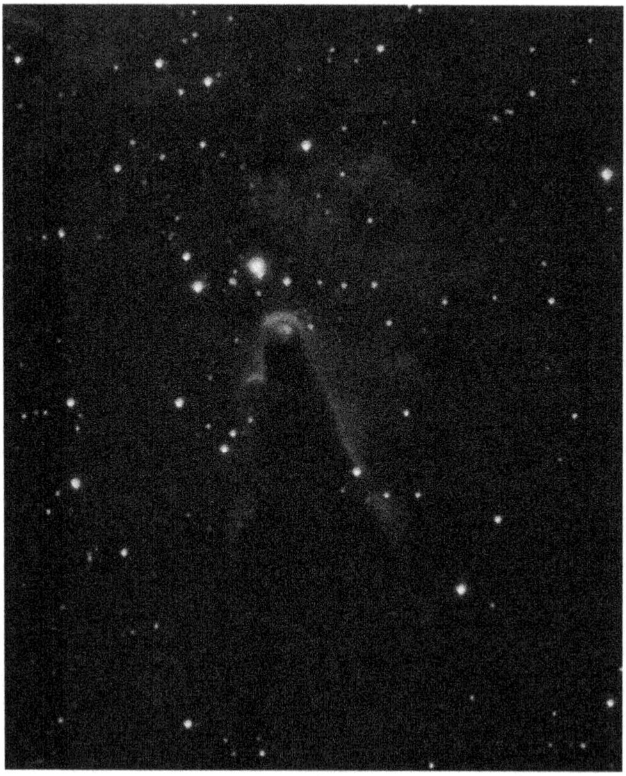

Fig. 8.36 NGC 2264

NGC 2362 Canis Major

RA: 07h 18m
Dec: −24.57
Size: 2.5×2′
Magnitude: 4.1
Filters: RGB

This is a fabulous little star cluster centered on τ Canis Majoris and is commonly known as the Northern Jewel Box due to the profusion of stars in this tiny area. Although the stars are hard to separate in a small telescope the cluster is an excellent target for imaging. Astronomical studies have revealed that the cluster contains 500 solar masses and is only about 5 million years in age and lies at a distance of 5,000 light years. Greater in extent and surrounding the cluster is the nebula Sharpless 2-310, which is a very difficult object from UK latitudes but would make a good target for southern observers. (Image taken by Martin Griffiths.)

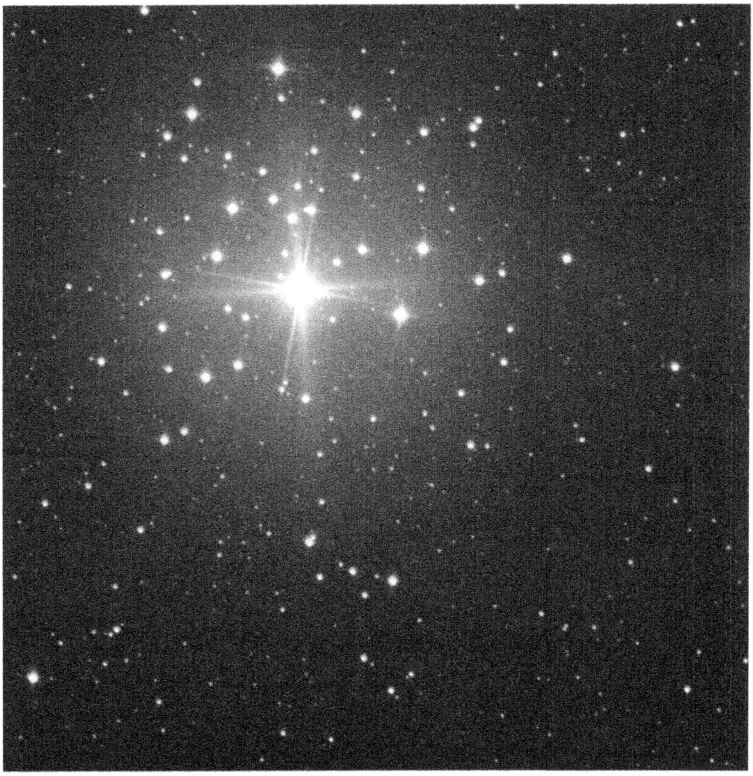

Fig. 8.37 NGC 2362

NGC 2392 Gemini

RA: 07h 29m
Dec: +20.54
Size: 4′
Magnitude: 10.2
Filters: Hα, OIII, SII, RGB

A superb planetary in southern Gemini, the "Eskimo Nebula" is beautifully green and immediately obvious no matter what sized telescope is used. It is an evenly illuminated disk at first glance, but under higher powers it easily separates into two distinct layers, a bright core with a slightly resolved central star and fuzzy outer layer representing the "hood" of the Eskimo. The hood should be readily apparent with averted vision, and the little face is almost there and can be glimpsed in high power with averted vision. NGC 2392 is a wonderful planetary, a favorite with many observers due to its magnitude, visibility and detail. Unsurprisingly, it is a Caldwell object (No 39) and Herschel 400 catalog object, too. (Image taken by Kenton Primary School with Faulkes Telescope North operated by Las Cumbres Observatory Global Telescope Network. Used with permission.)

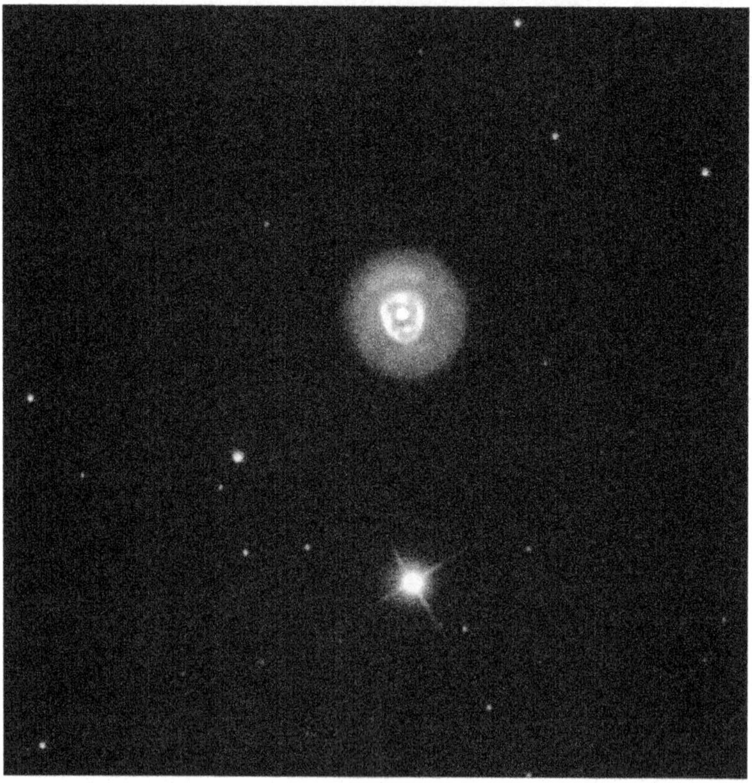

Fig. 8.38 NGC 2392

NGC 2437/38 (Messier 46) Puppis

RA: 07h 241m
Dec: −14.49
Size: 27×24′
Magnitude: 6.1
Filters: Hα, OIII, SII, RGB

This is the star cluster M46, with the apparently embedded planetary nebula NGC 2438. The stars will reveal themselves as a nice compacted object, but the observer should attempt to draw out the blue colored planetary on its northern edge. In 1963 the astronomer C. R. O'Dell determined that the association is merely a line of sight effect, and that the radial velocity of the cluster and that of the nebula are quite different; the nebula appears to be a foreground object, but the juxtaposition of the two objects is quite pleasing. (Image taken by Angela Keen with Faulkes Telescope North operated by Las Cumbres Observatory Global Telescope Network. Used with permission.)

Fig. 8.39 NGC 2437/38

NGC 2403 Camelopardalis

RA: 07h 36m
Dec: +65.36
Size: 22 × 12′
Magnitude: 8.9
Filters: Hα, NB, RGB

This is a fantastic large galaxy in Camelopardalis that is part of the Messier 81 group, yet in contrast is only 8 million light years away in comparison to M81's 12 million light years. NGC 2403 reveals a mottled appearance with a large nucleus and ill-defined arms that are ablaze with HII regions and spread into feathery outliers. The whole object is glowing with newly formed stars, and it is possible that it has experienced some kind of merger in the recent past, as much of its gas content is actually counter rotating in comparison to the flow of the spiral arms. It lies at a slight angle to our line of sight, too, which makes it a large ovoid object that should fill the field on a CCD. Several supernovae have been observed in this galaxy over the past century, so it is a good idea to image it frequently with the hope of discovering one of these elusive events. (Image taken by Hull Collegiate School with Faulkes Telescope North operated by Las Cumbres Observatory Global Telescope Network. Used with permission.)

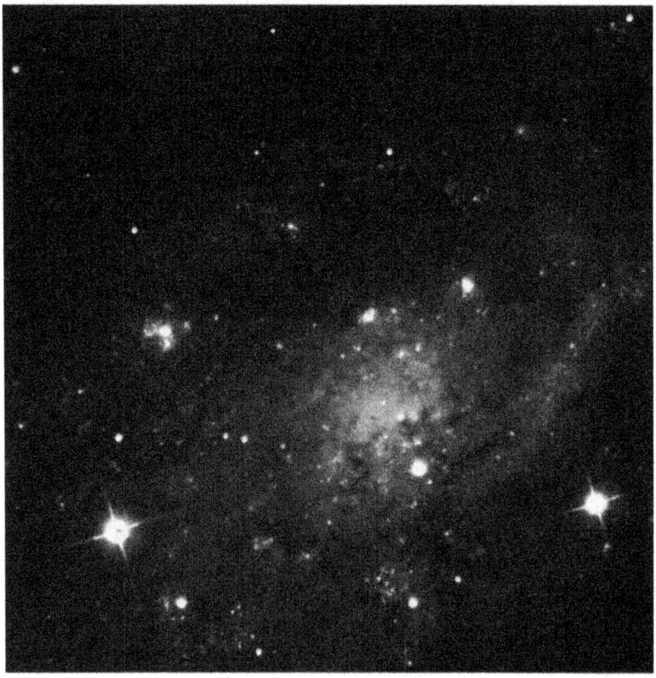

Fig. 8.40 NGC 2403

NGC 2440 Puppis

RA: 07h 41m
Dec: −18.12.5
Size: 74″
Magnitude: 9.4
Filters: Hα, OIII, SII, RGB

Almost directly south of M46 is this cuboid of white light containing a white dwarf star, which is one of the hottest known. The central star is not visible however due to a dense central concentration, but the nebula does reveal a lovely bipolar form oriented on an E–W axis. In a higher power there appears a nice "bite"-type bay on its N side. There is little difficulty in seeing structure in most telescopes, as the nebula is quite bright at around mag 9.5 and is over 1′ in extent across the main axis. No obvious color is visible apart from the gray-white of the nebula. NGC 2440 is also a Herschel 400 object. (Image taken by St George's School with Faulkes Telescope North operated by Las Cumbres Observatory Global Telescope Network. Used with permission.)

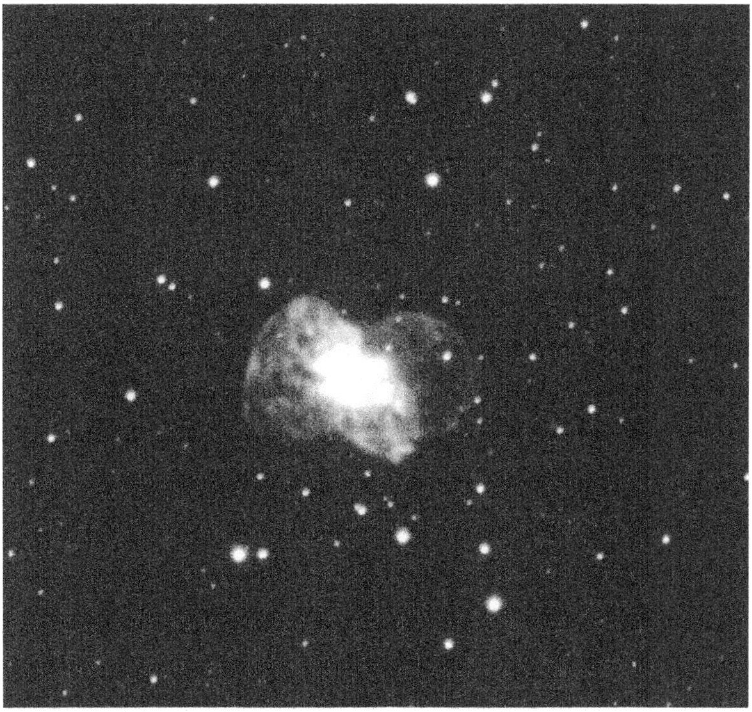

Fig. 8.41 NGC 2440

IC 1396 Cepheus

RA: 08h 37m
Dec: +56.45
Size: 45×40′
Magnitude: 9.1
Filters: Hα, NB, RGB

This is an enormous cloud of hydrogen gas illuminated by a series of star clusters and newborn stars that is over 100 light years across and decorates the southern edge of the constellation Cepheus. One of the most photographed parts of this large nebula is the "elephant's trunk," which intrudes into the center of this almost circular object and is an area of recent star formation. The rest of this red nebula spreads over an area twice the width of the Moon and is spattered with dark globules, dust lanes and bright ionization fronts and reveals a curious dark "hole" just off center of the nebula. It is easily located just to the south of μ Cephei—Herschel's "garnet star"—and is a circumpolar object for northern observers living above 40° latitude. (Image taken by Martin Griffiths.)

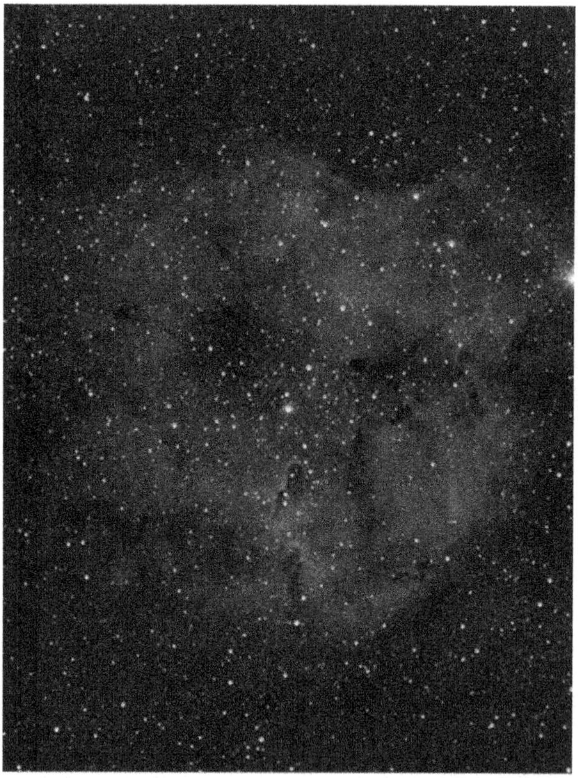

Fig. 8.42 IC 1396

NGC 2818 Pyxis

RA: 09h 16m
Dec: −36.37
Size: 4′
Magnitude: 5.6
Filters: Hα, OIII, SII, RGB

This nebula is part of a lovely star cluster that is immediately visible in most telescopes, and the planetary is superimposed on it as a gray, elongated patch of light about tenth magnitude and 40″ or more in extent along an E–W axis surrounded by the lovely blue-white stars of the cluster. The planetary nebula is on the NW side of the group and is obvious in low power as a whitish patch of light. Higher magnifications render it more detailed, with a dark bay on each side of the dense central portion. Although it is apparent with direct vision, the nebula shows a good response to an OIII filter. (Image taken by Ysgol Glan Clwyd with Faulkes Telescope North operated by Las Cumbres Observatory Global Telescope Network. Used with permission.)

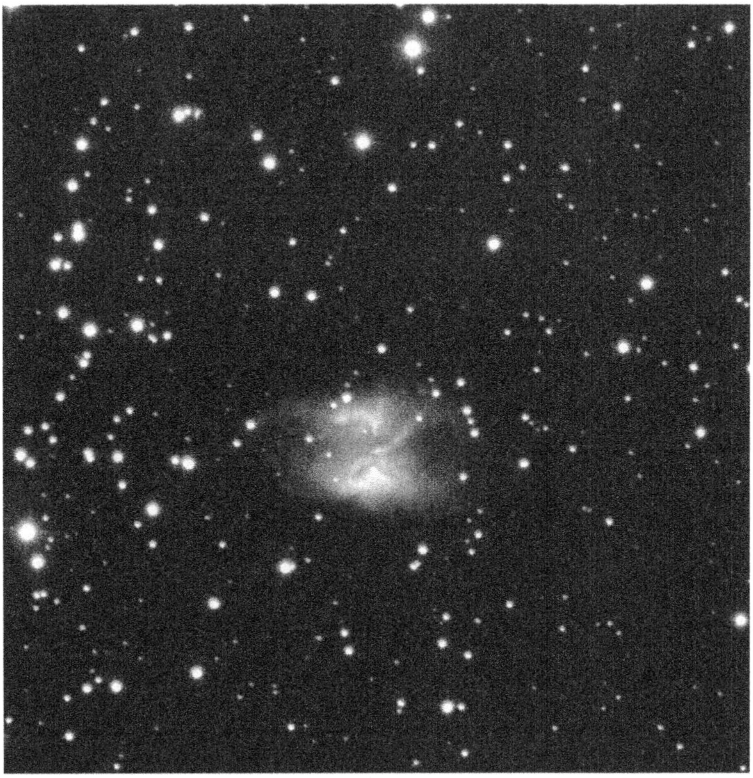

Fig. 8.43 NGC 2818

NGC 2841 Ursa Major

RA: 09h 22m
Dec: +50.58
Size: 8×4′
Magnitude: 10.1
Filter: RGB

This is a lovely little spiral galaxy in Ursa Major that, despite its distance of 46 million light years, is still a delightful object revealing a lot of detail in both telescope and image. The galaxy is quite flocculent and has a wealth of tiny HII regions in the arms, small dark dust patches across its surface and many star clouds, which all make a lovely tight winding motion around its starry nucleus. In actual size it is a very large spiral galaxy—over 150,000 light years across, almost half the size again of our own Milky Way. (Image taken by Light Hall School with Faulkes Telescope North operated by Las Cumbres Observatory Global Telescope Network. Used with permission.)

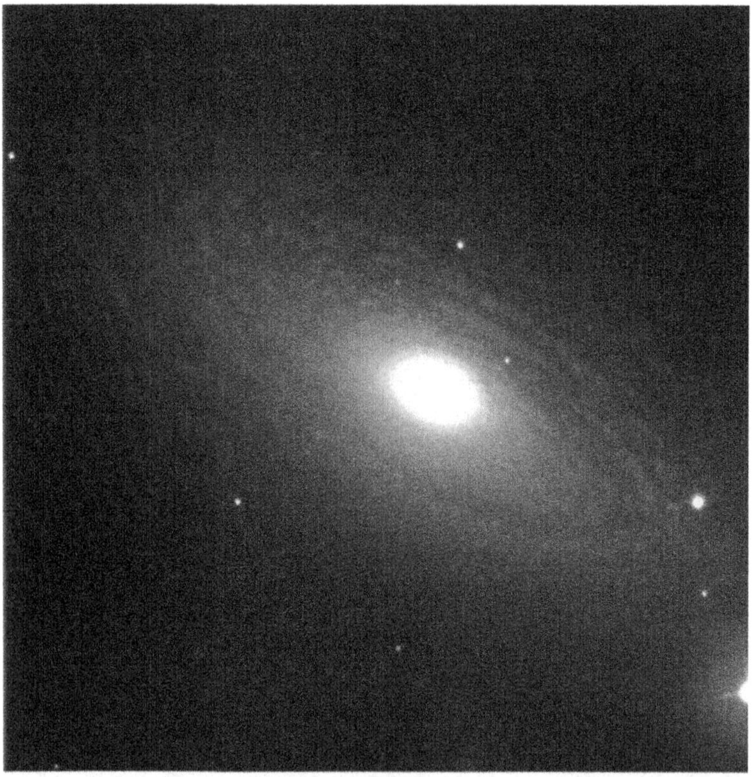

Fig. 8.44 NGC 2841

NGC 2903 Leo

RA: 09h 32m
Dec: +21.30
Size: 12×6′
Magnitude: 8.4
Filters: Hα, NB, RGB

A lovely large and extended barred spiral galaxy in Leo just off the top of the lion's mane, NGC 2903 is over 30 million light years away, inclined to our line of sight and has a wonderful swirl of spiral arms that become distended and feathery at their ends. It is a bright object that, for unknown reasons, Messier and a host of other observers missed until it was spotted by William Herschel in 1784. It has a bright core region, and the arms separate out at the junction of two dust lanes in the redder interior of the galaxy and terminate in a trail of blue stars. It has undergone a recent starburst in a ring about 2,000 light years from the galactic center—a feature that may show up with appropriate editing in *Photoshop*. (Image taken by Faringdon Community College with Faulkes Telescope North operated by Las Cumbres Observatory Global Telescope Network. Used with permission.)

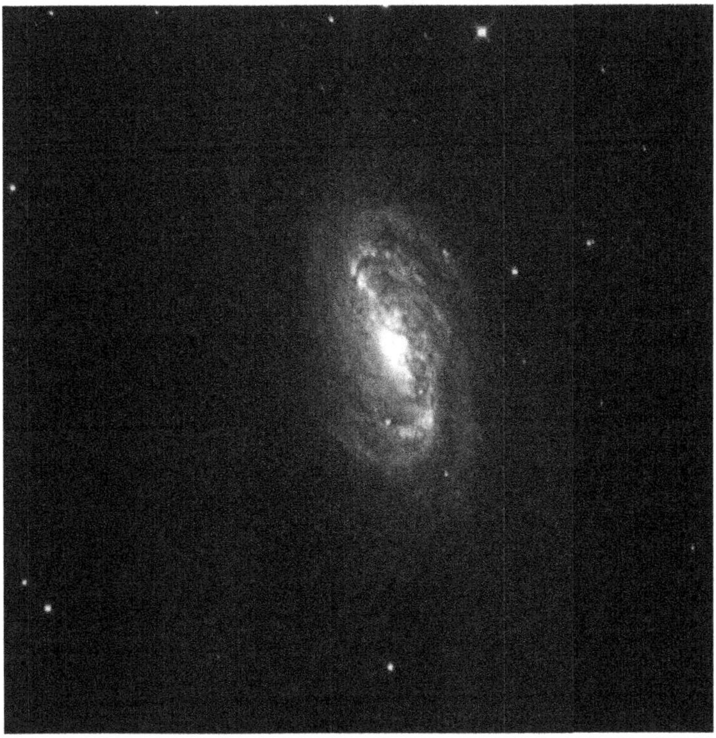

Fig. 8.45 NGC 2903

Messier 82 Ursa Major

RA: 09h 55m
Dec: +69.40
Size: 12×4′
Magnitude: 8.4
Filters: Hα, NB, RGB

The beautiful "Cigar Galaxy" in Ursa Major is a favorite target for many photographers. M82 is a starburst galaxy 12 million light years away that has undergone interaction with its larger neighbor (M81) and has almost blown up in its excitement—as can be seen from the lanes of dust and energized gas that cross its center and extend out into space. The electric blue color of its spiral edge-on structure and the contrast with the redness of its gaseous lanes make this galaxy an excellent target for all astronomers. (Image taken by Institute for Astronomy Maui, Hawaii with Faulkes Telescope North operated by Las Cumbres Observatory Global Telescope Network. Used with permission.)

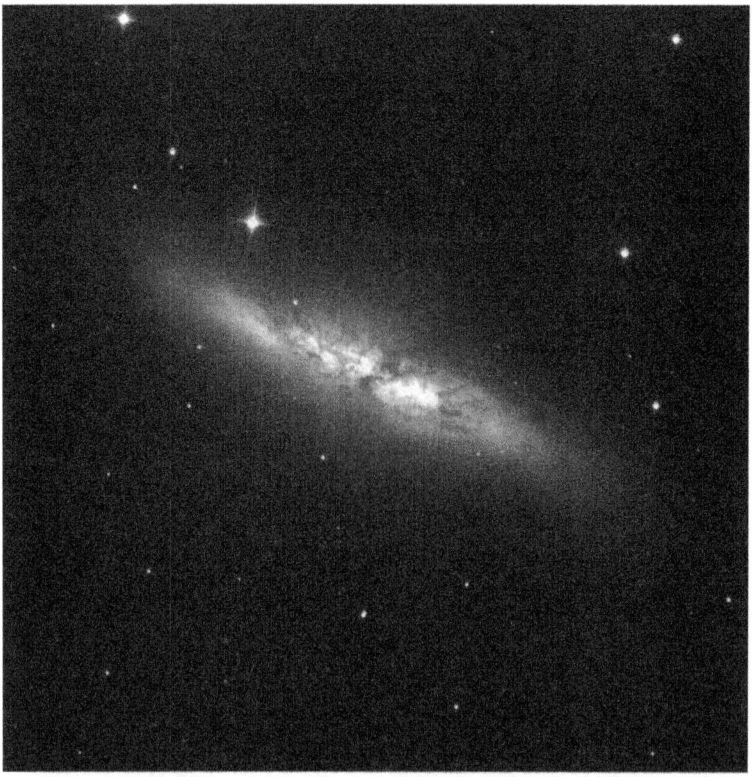

Fig. 8.46 M82

NGC 3132 Vela

RA: 10h 07m
Dec: −40.26.0
Size: 60″
Magnitude: 9.7
Filters: Hα, OIII, SII, RGB

This beautiful planetary, known as the "Eight Burst Nebula" or the "Southern Ring Nebula," is an easy object even in small telescopes. In low powers it is obviously elongated, with a SE/NW axis, has a thick edge all around its cuboid shape and has a filled core revealing (with averted vision) a central star. The interior is a gray-white with a hint of blue, but the outer edges seem definitely red in tinge (very much like M57's color) with a hint of purple in averted vision at moderate magnification. A wonderful object for all telescopes. (Image taken by Martin Griffiths with Faulkes Telescope North operated by Las Cumbres Observatory Global Telescope Network.)

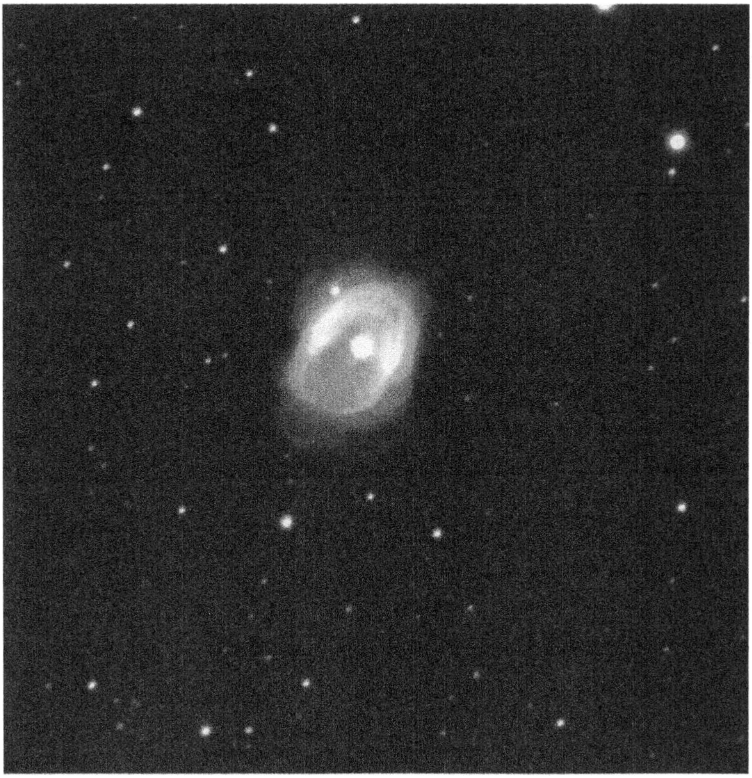

Fig. 8.47 NGC 3132

NGC 3195 Chameleon

RA: 10h 09m
Dec: −80.55.11
Size: 35″
Magnitude: 10.1
Filters: Hα, OIII, SII, RGB

This wonderfully bright oval nebula is an easy target for modestly equipped observers and is very high in the southern circumpolar sky. NGC 3195 is a bright filled shell at low powers, which stands revealed as a planetary that some observers have described as rather like M97 with two prominent dark bays at higher magnifications and use of averted vision. At high powers the object is mottled in appearance and is evenly filled with a greenish light. NGC 3195 can be found midway between ζ and δ Chameleon and responds well to OIII filters, though no central star is to be seen. (Image taken by Old Palace School with Faulkes Telescope North operated by Las Cumbres Observatory Global Telescope Network. Used with permission.)

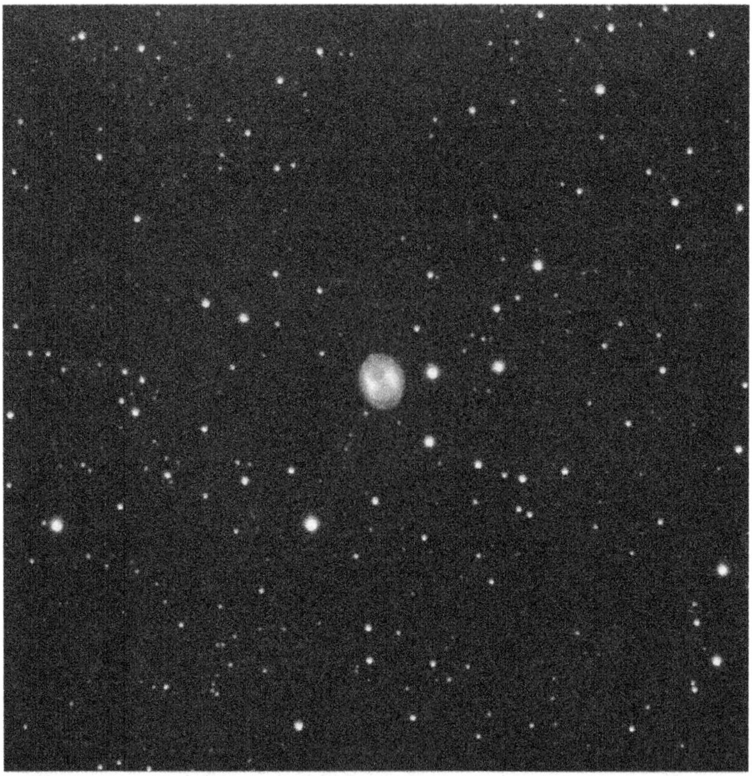

Fig. 8.48 NGC 3195

NGC 3242 Hydra

RA: 10h 26m
Dec: −18.38.5
Size: 45″
Magnitude: 7.8
Filters: Hα, OIII, SII, RGB

The wonderful "Ghost of Jupiter Nebula" is an easy target south of μ Hydra and reveals itself as a round filled shell in a lovely aquamarine color. It has some slight ellipsoid structure with an E–W axis. It is almost 1′ across, and at only eighth magnitude it is visible in binoculars and definitely visible in telescopes of most apertures. NGC 3242 is in a rather lonely field and thus stands out rather well. This object is also known as Caldwell 59 and is also in the Herschel 400 catalog. NGC 3242 remains one of the brightest and best planetaries in the sky, though it was sadly overlooked by Messier and his observing colleagues. (Image taken by Martin Griffiths with Faulkes Telescope North operated by Las Cumbres Observatory Global Telescope Network.)

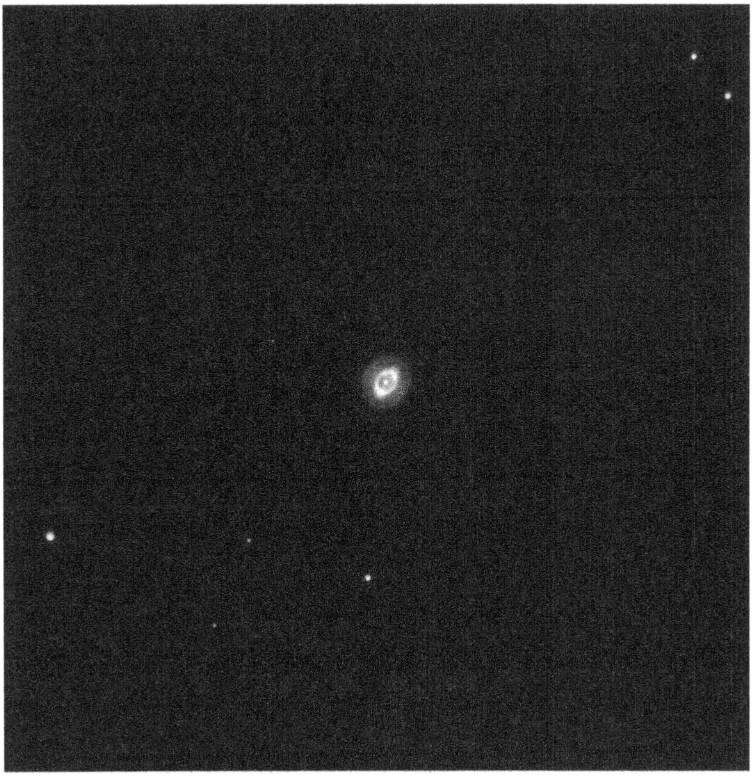

Fig. 8.49 NGC 3242

NGC 3372 Carina

RA: 10h 45m
Dec: −59.52
Size: 50×45′
Magnitude: 2.8
Filters: Hα, NB, RGB

The huge complex known as the "Eta Carina Nebula" is well known to southern hemisphere astronomers and makes a very tempting target with its huge size and so many objects that are available for exploring within it. The Homunculus Nebula, the Mystic Mountain Nebula and the Keyhole Nebula (NGC 3324) are the obvious candidates to separate out within this complex, and all are lit by O- and B-type stars in an enormous association that spreads across 50 light years at the distance of 6,000 light years to the main parts of this nebula. The beautiful variations of color seen in observatory pictures can be captured with Hα and SI filters as well as the usual broadband nebula and color filters. A visually stunning object and an astrophotographers dream. (Image taken by South Downs Planetarium with Faulkes Telescope South operated by Las Cumbres Observatory Global Telescope Network. Used with permission.)

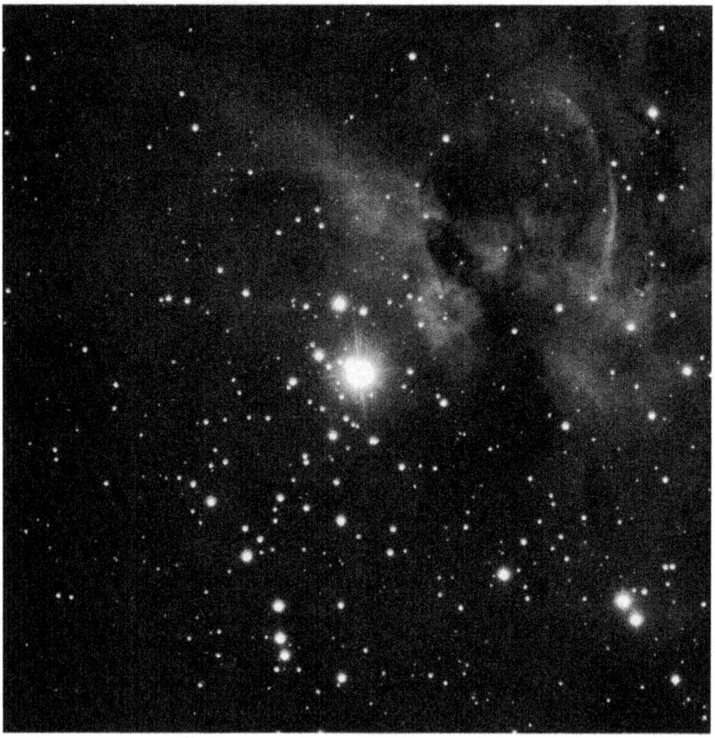

Fig. 8.50 NGC 3372

Messier 97 Ursa Major

RA: 11h 14m
Dec: +55.01.01
Size: 200″
Magnitude: 9.8
Filters: Hα, OIII, SII, RGB

This is the fabulous "Owl Nebula" in the Great Bear, which shares the same low power field as the galaxy NGC 3556 and provides a nice contrast between two Messier objects (97 and 108). M97 is a wonderfully detailed object in most telescopes, where it is visible as a ghostly greenish glow against the background stars. It is large and round, while higher magnifications on larger telescopes bring out the two dark bays toward the center and may on good nights reveal the central star. This is a wonderful planetary nebula and always visible in the circumpolar skies of the northern hemisphere. (Image taken by The Burgate School with Faulkes Telescope North operated by Las Cumbres Observatory Global Telescope Network. Used with permission.)

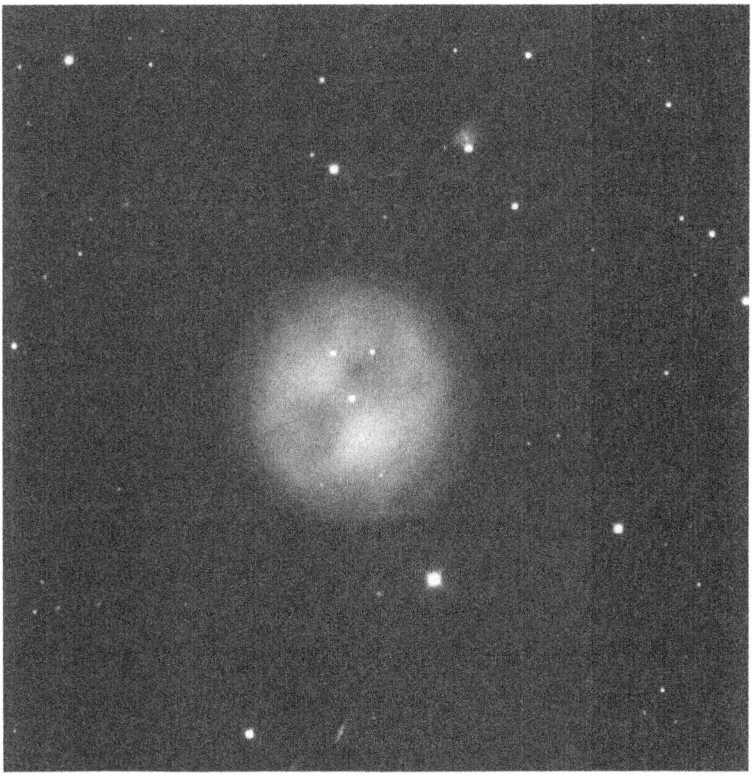

Fig. 8.51 M97

NGC 3628 Leo

RA: 11h 20m
Dec: +18.35
Size: 15×3′
Magnitude: 9.9
Filters: Hα, NB, RGB

The wonderful "Hamburger" Galaxy in Leo, forming part of the so-called "Leo triplet" with M65 and 66. Lying about 30 million light years away it has an instantly recognizable dust lane that cuts through its center, and fluffy star clouds outline the rest of the galaxy lying above and below this lane that give rise to its characteristic name. In total extent the galaxy is over 300,000 light years across, as it has interacted with M65, leading to a long tidal "tail." Opposite this tail the dust lane diverges into several components, radiating out until lost against the darkness of space. All three are a fantastic trio to image and a low f number telescope may enable the observer to capture them all in one field. (Image taken by the Lockerbie Academy with Faulkes Telescope North operated by Las Cumbres Observatory Global Telescope Network. Used with permission.)

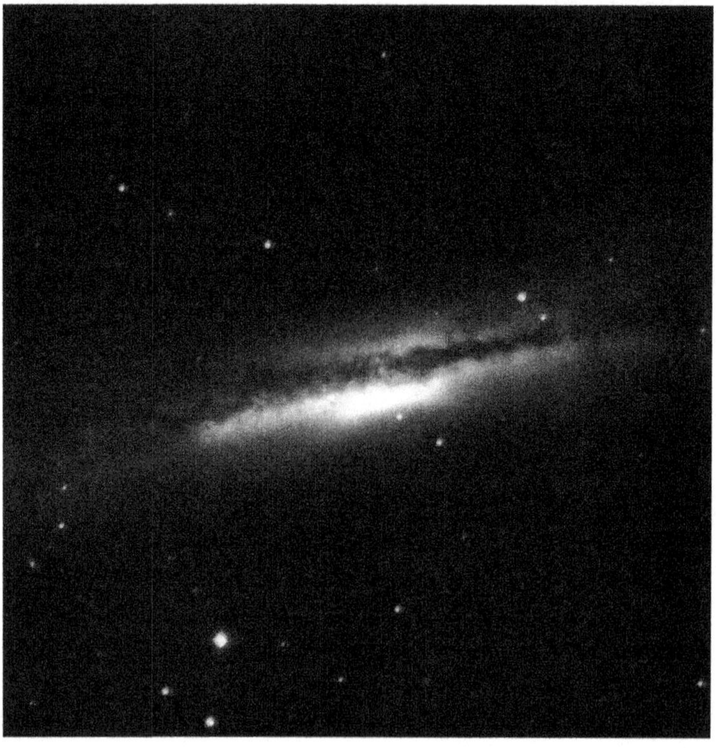

Fig. 8.52 NGC 3628

Messier 106 Canes Venatici

RA: 12h 18m
Dec: +47.18
Size: 19×7′
Magnitude: 9.1
Filters: Hα, NB, RGB

This is another great galaxy and one undergoing internal interactions—possibly with a super massive black hole at the center. Its overall appearance is as a tortured spiral with many overlaid red gas and dust lanes leading into the center from well-defined spiral arms that show many signs of starburst activity along their axes. Strong Hα emitter, and with so many nebulae across the galaxy an Hα filter will bring out all the possibilities of this lovely object. Messier 106 is over 23 million light years away and may also be interacting with a companion galaxy NGC 4217. (Image taken by Olsztsyn School with Faulkes Telescope North operated by Las Cumbres Observatory Global Telescope Network. Used with permission.)

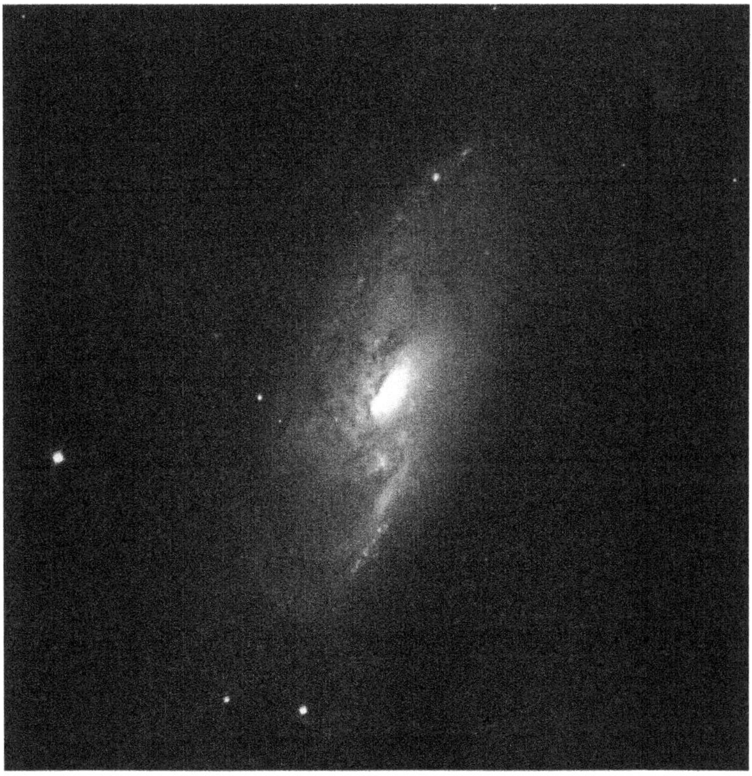

Fig. 8.53 M106

Messier 100 Coma Berenices

RA: 12h 22m
Dec: +15.49
Size: 7×6′
Magnitude: 10.1
Filters: Hα, NB, RGB

The great "Grand Design" spiral galaxy in Coma Berenices lies at 55 million light years away as a face-on galaxy, with arms revealing extraordinary detail as they swirl out from the small nucleus into a star and HII studded expansion of light and dust lanes. Larger than the Milky Way at 160,000 light years across it is a favorite target for amateur astronomers and was used by various academic groups to gauge the Hubble constant back in the 1990s. Since the beginning of the twentieth century, five supernova have been recorded in this galaxy, and its large size and flat appearance make it a lovely object for capture by CCD with its varied colors from starbursts, dust lanes and HII regions. (Image taken by South Downs Planetarium with Faulkes Telescope North operated by Las Cumbres Observatory Global Telescope Network. Used with permission.)

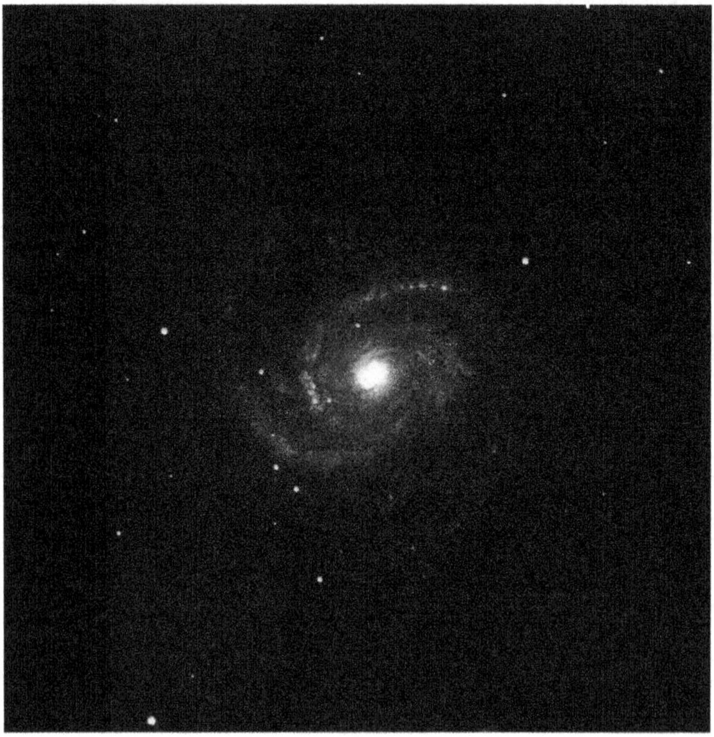

Fig. 8.54 M100

NGC 4361 Corvus

RA: 12h 24m
Dec: −18.47.1
Size: 70″
Magnitude: 10.9
Filters: Hα, OIII, SII, RGB

This is a lovely object that is easy to find as the peak of a scalene triangle from δ Corvi. The nebula is immediately apparent as a dual-lobed shell around a faint star in low powers, while higher magnifications increase the separation of star and nebula a little more, filling in the concentric gas cloud around the star. The nebula is blue-gray in color and remains that hue no matter what magnification is used. A pleasing object though often overlooked for its more obvious neighbor NGC 3242 in Hydra. It's also a Herschel 400 catalog object and quite a lovely planetary in its own right. (Image taken by Martin Griffiths with Faulkes Telescope North operated by Las Cumbres Observatory Global Telescope Network.)

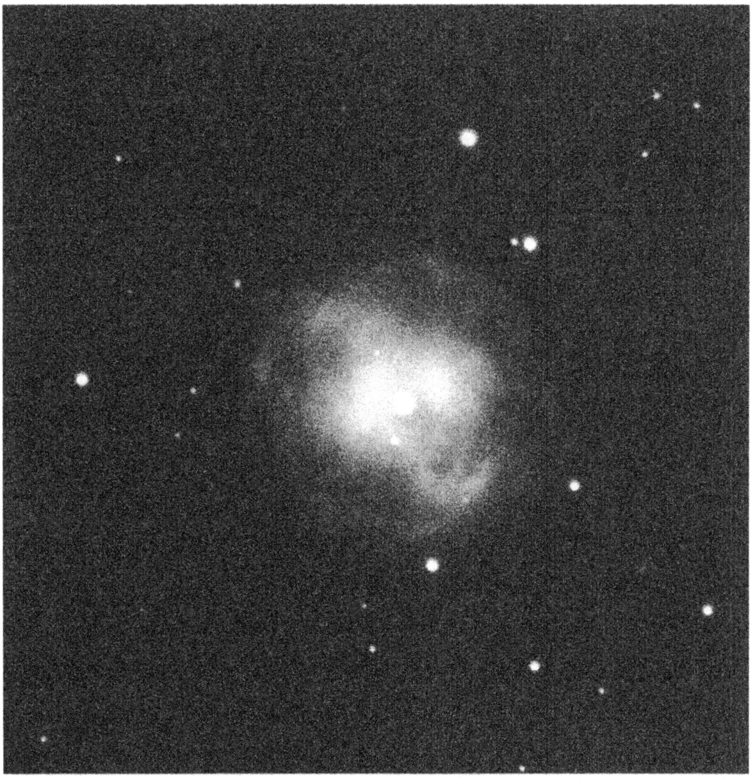

Fig. 8.55 NGC 4361

NGC 4485 Canes Venatici

RA: 12h 30m
Dec: +41.42
Size: 4×2′
Magnitude: 9.7
Filters: Hα, NB, RGB

This is the famous "Cocoon Galaxy" in Canes Venatici, which is interacting with its partner NGC 4490. Both galaxies are visible in a small telescope despite their reported distance of 42 million light years due to the fact that both are undergoing starbursts after their collision and both are distorted objects, with NGC 4485 having a pronounced teardrop shape with the thin end trailing into space in the direction of NGC 4490. The Cocoon Galaxy is mottled with knots of clusters and HII regions, and good imaging will also bring out the knots of starlight that lie in the bridge of material that tie the two galaxies together. Relatively easy to find just to the NW of β Canes Venaticorum. (Image taken by Old Palace School with Faulkes Telescope North operated by Las Cumbres Observatory Global Telescope Network. Used with permission.)

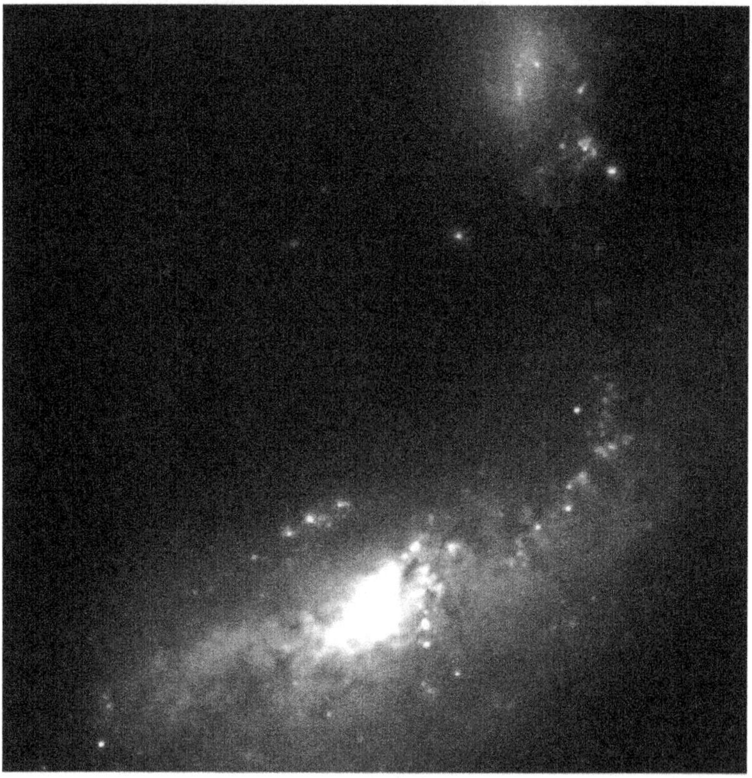

Fig. 8.56 NGC 4485

Messier 87 Virgo

RA: 12h 30m
Dec: +12.23
Size: 8×6′
Magnitude: 9.5
Filters: Hα, RGB

This is one of the most massive galaxies known and is surrounded by a hive of globular clusters. A giant elliptical galaxy in Virgo, it is 53 million light years away and is visible in binoculars as a dull round spot of light. The one feature that many astrophotographers attempt to capture is the tiny beam of material and light emanating from the center from a super massive black hole. This jet of electrons and protons is moving at a considerable velocity and is close to the center of the galaxy—easily swamped by overexposure if the observer is not careful. In reality, the jet is over 5,000 light years in length—testament to the underlying power of the engine at the center of this galaxy. The jet is a challenging object but makes recording this galaxy photographically a good challenge for the enthusiast. (Image taken by Glenlola Collegiate with Faulkes Telescope North operated by Las Cumbres Observatory Global Telescope Network. Used with permission.)

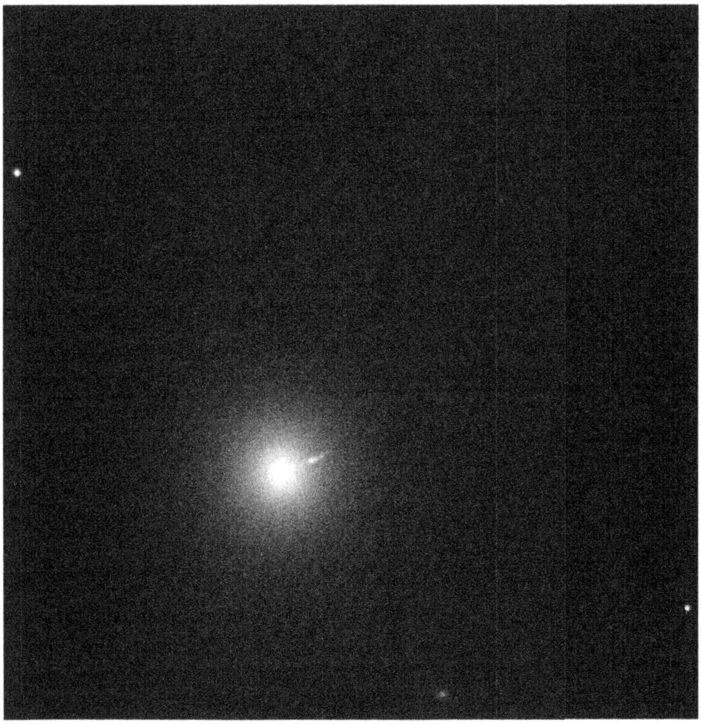

Fig. 8.57 M87

NGC 4565 Coma Berenices

RA: 12h 36m
Dec: +25.59
Size: 15 × 2′
Magnitude: 10.4
Filters: Hα, RGB

This is one of the most beautiful galaxies in the entire sky and a favorite target for observers and photographers alike. The "Needle Galaxy," NGC 4565, is a spiral that is perfectly edge-on to our line of sight and displays a wonderful thin dust lane along its entire length, with a round nucleus symmetrically placed at the center. The galaxy has two companions that are interacting with it, but they are difficult objects to image. In the same field of view is the more distant galaxy NGC 4562. The Needle Galaxy lies at a distance of 45 million light years, and recent studies have revealed that this object is actually a barred spiral. If it was face-on to us it would be as fantastic an object as M83. (Image taken by Martin Griffiths with Faulkes Telescope North operated by Las Cumbres Observatory Global Telescope Network.)

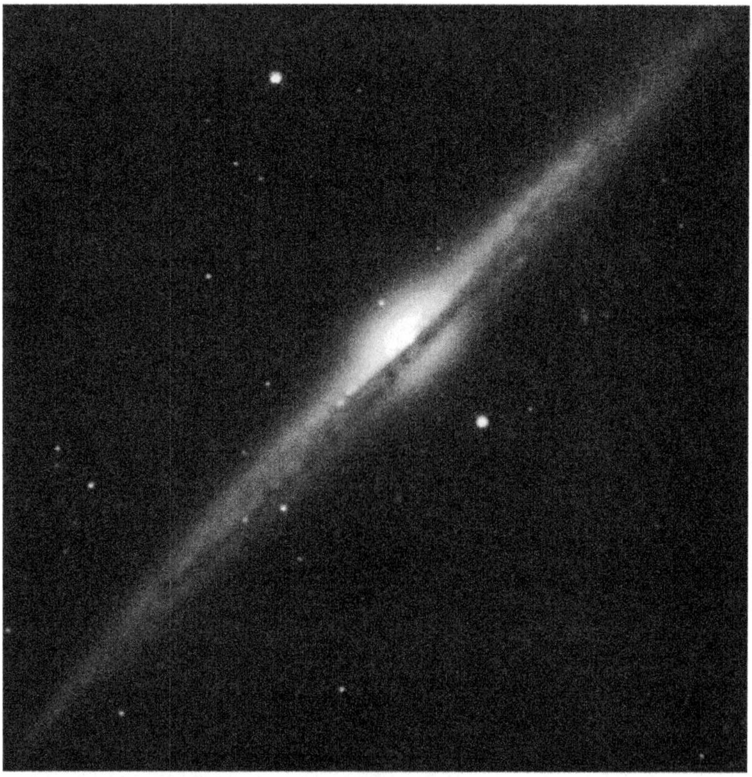

Fig. 8.58 NGC 4565

Messier 104 Virgo

RA: 12h 39
Dec: −11.37
Size: 9×3′
Magnitude: 8.9
Filters: Hα, RGB

Another spectacular galaxy, the "Sombrero Galaxy" lies on the Virgo/Corvus border about 28 million light years away from us and revealing a wonderful lane of dust set against a glowing disc of stars with a prominent nucleus. Long-exposure photographs show a large halo of stars eclipsing the nucleus of this SO type galaxy and ruining the "sombrero" effect somewhat. Despite its proximity to the Virgo cluster M104 is not actually a member of that super cluster but is part of a small cloud of galaxies with different radial velocities to the main Virgo group. Some small HII regions dot the dust lanes, but the lane itself is the superlative feature of this lovely galaxy. (Image taken by South Downs Planetarium with Faulkes Telescope North operated by Las Cumbres Observatory Global Telescope Network. Used with permission.)

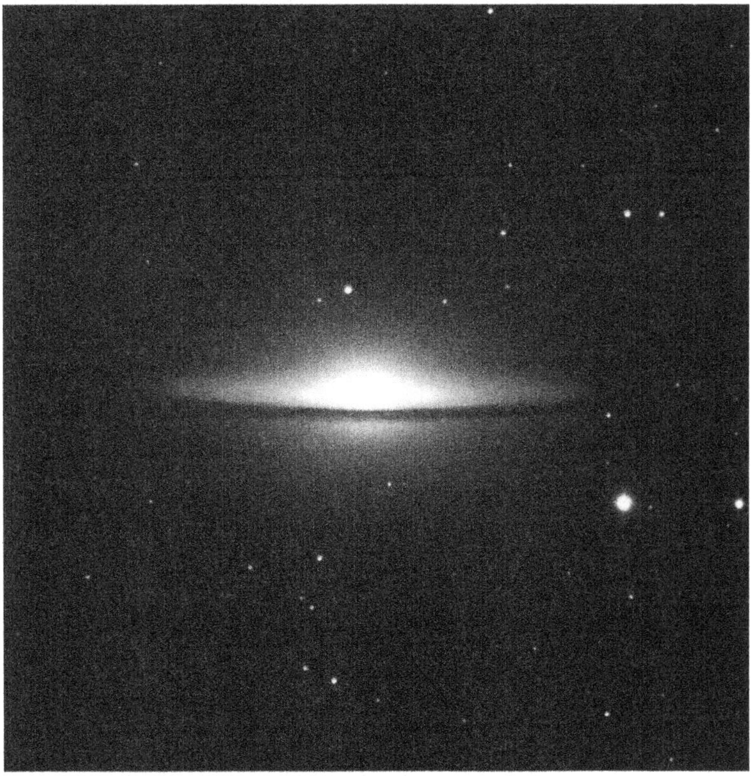

Fig. 8.59 M104

NGC 4631 Canes Venatici

RA: 12h 42m
Dec: +32.32
Size: 15′×2.5′
Magnitude: 9.8
Filters: Hα, RGB

The wonderful "Whale Galaxy" is an edge-on strip of light that complements NGC 4656 in the field of view of a small telescope. Imaging the galaxy brings out the vivid blue hues that delineate star formation along its length and terminate in a huge burst of starlight in a small tail at the western end of the object. The central dust lane of the galaxy is almost overwhelmed, so careful imaging is necessary, but since the object is bright, obtaining images should not be a problem. Look for the tiny elliptical companion above its main body. NGC 4631 is 30 million light years away. (Image taken by Martin Griffiths.)

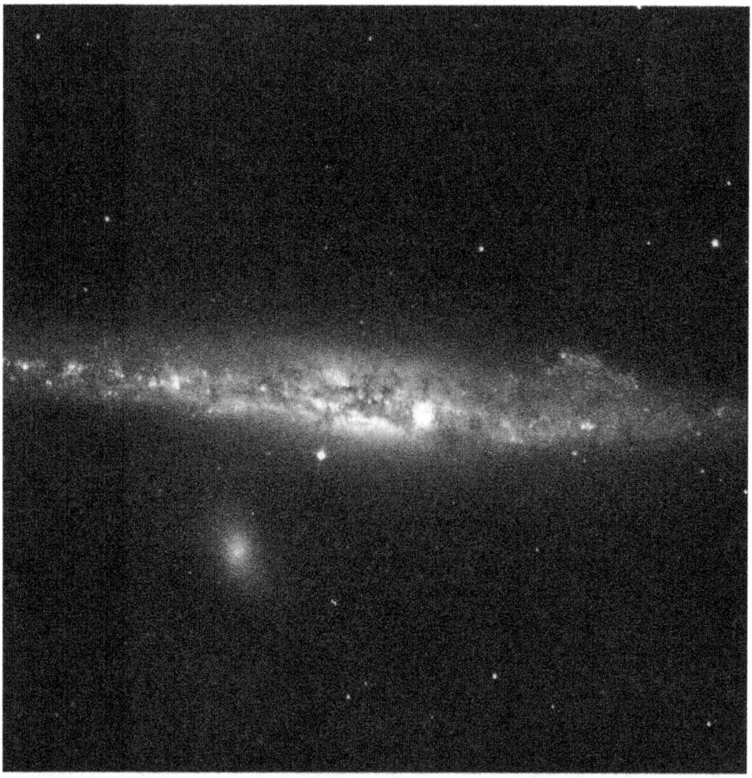

Fig. 8.60 NGC 4631

NGC 4656 Canes Venatici

RA: 12h 43m
Dec: +32.10
Size: 10×3′
Magnitude: 11.1
Filters: Hα, RGB

The "Hockey Stick Galaxy" in Canes Venatici makes a pleasing pair with NGC 4631 and is extremely blue and mottled with star clouds due to recent interaction and starburst activity. The galaxy lies about 30 million light years away, and the twisted end of one of its arms is evident even in short exposures and can be explored visually with a 15-cm reflector where an off center nucleus can be discerned. In color photographs this reveals itself as a reddish patch of light in contrast to the intense blue of the arms of the galaxy. HII regions dot the arms, but fine detail can be difficult to pull out unless the field of view of the CCD is small enough to contain only NGC 4565. (Image taken by Olsztyn School with Faulkes Telescope North operated by Las Cumbres Observatory Global Telescope Network. Used with permission.)

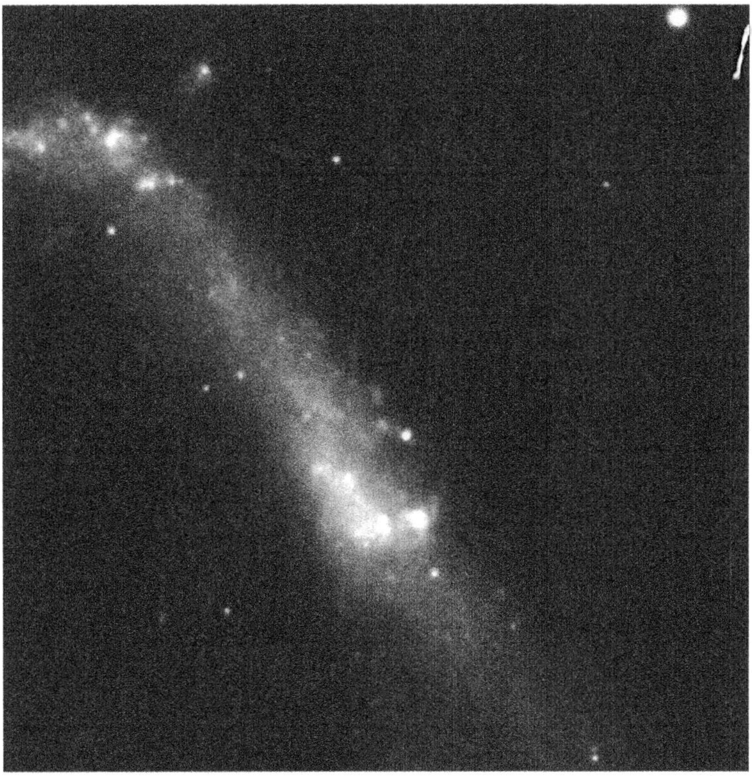

Fig. 8.61 NGC 4656

NGC 4755 Crux

RA: 12h 53m
Dec: −60.22
Size: 15 × 15′
Magnitude: 4.1
Filter: RGB

The "Jewel Box" cluster, so named by John Herschel is one of the finest sights in the southern sky and a worthy target for any observer. The cluster contains a red giant at the center that makes a pleasing contrast to the blue stars of the rest of the group, which number over 100 in total. At 6,400 light years away the concentration of this group is due to distance—which is probably why it has the Bayer designation κ Crucis, since, to the naked eye, it looks like a single star. It is one of the youngest known open clusters, at less than 10 million years old, and is a beautiful object that observers return to time and again. (Image taken by Martin Griffiths with Faulkes Telescope South operated by Las Cumbres Observatory Global Telescope Network.)

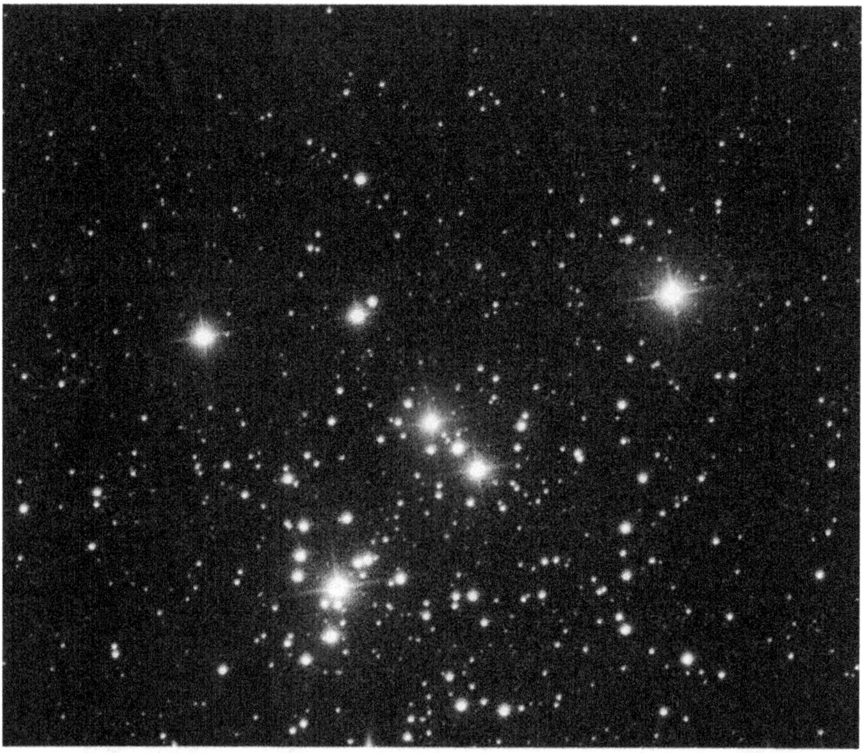

Fig. 8.62 NGC 4755

NGC 5053 Coma Berenices

RA: 13h 16m
Dec: +17.42
Size: 50×50′
Magnitude: 9.4
Filter: RGB

A faint and visually difficult globular cluster, this is only a degree or so away from the globular cluster M53 in Coma Berenices. NGC 5053 is almost ‘transparent,’ as it does not have the usual concentration common to such objects but remains a scattered smattering of starry points about 10′ across but with little in the way of magnitude differences or concentration at all. Still, as such an unusual object it poses a challenge for observers and photographers. The cluster contains many RR Lyra-type Cepheid variables, which hint at its extreme age. (Image taken by Martin Griffiths with Faulkes Telescope North operated by Las Cumbres Observatory Global Telescope Network.)

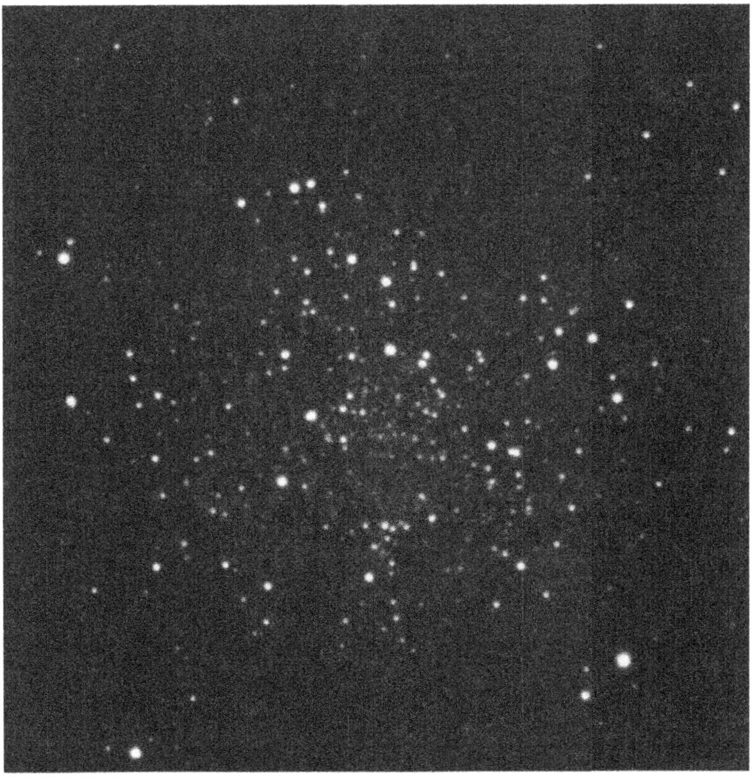

Fig. 8.63 NGC 5053

NGC 5128 Centaurus

RA: 13h 25m
Dec: −43.01
Size: 21 × 19′
Magnitude: 6.8
Filters: Hα, NB, RGB

The wonderful elliptical galaxy Centaurus A is a radio source, but for most amateurs it is the lure of recording the fantastic dust lane that crosses this interacting galaxy that makes it such a splendid sight. NGC 5128 lies about 14 million light years away and is the result of a galaxy merger between an elliptical and a small spiral galaxy. Starbursts across the galaxy are still ongoing, and materials have fallen onto the central super massive black hole awakening it to all sorts of activity, producing two huge radio jets, blasting out of the galaxy in a similar mechanism to the jet visible in M87. In 1988 a bright supernova was recorded here, made more brilliant by its backdrop against the central dust lane that is the most obvious component of this object. (Image taken by Martin Griffiths with Faulkes Telescope South operated by Las Cumbres Observatory Global Telescope Network.)

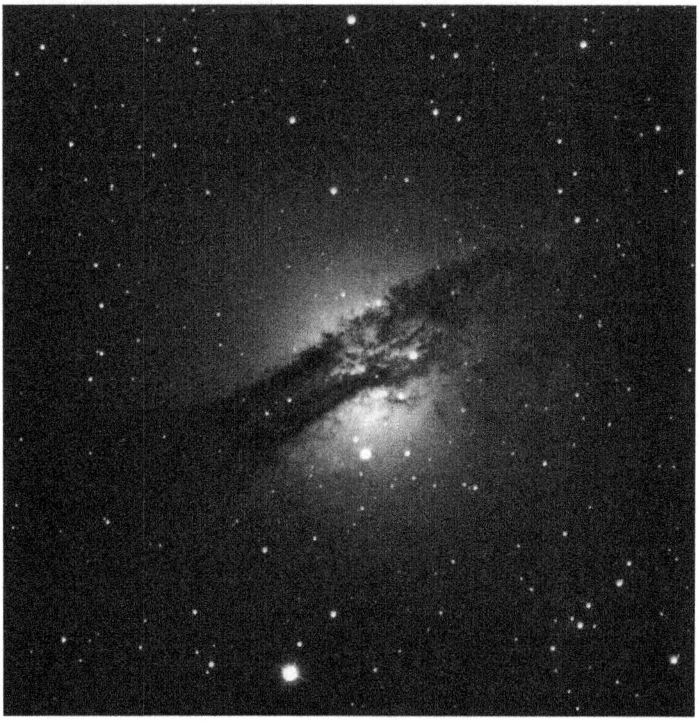

Fig. 8.64 NGC 5128

NGC 5139 Centaurus

RA: 13h 26m
Dec: −47.28
Size: 55×60′
Magnitude: 3.7
Filter: RGB

Omega Centauri is one of the greatest globular clusters in the sky and is a wonderful object to image, revealing a wealth of detail even in short exposures. It is the brightest and most massive globular cluster associated with the Milky Way and lies at a distance of 15,000 light years. Due to its sheer size and mass the cluster has been suggested to be the disrupted center of an interacting dwarf galaxy, and its metallicity and range of stars seem to confirm this to be the case. NGC 5139 shows up well in DSLR shots as well as in CCD cameras. (Image taken by St Catherine's College with Faulkes Telescope South operated by Las Cumbres Observatory Global Telescope Network. Used with permission.)

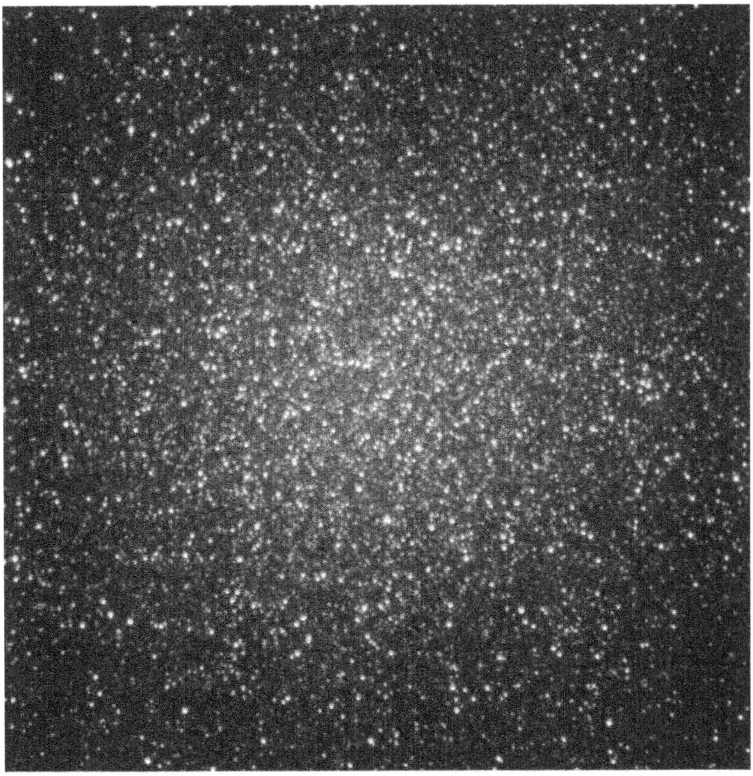

Fig. 8.65 NGC 5139

NGC 5189 Musca

RA: 13h 33m
Dec: −65.58.2
Size: 90″
Magnitude: 10.1
Filters: Hα, OIII, SII, RGB

This is a large amorphous body that is often marked on star charts as a small nebula rather than a planetary, though it is included in the P–K catalog and is an example of an emission nebula with Hα excess and a distended object that can mimic aged planetary nebula. In low power the nebula is gray-blue in color, elongated on a NE/SW axis and in higher magnifications on larger telescopes and exhibits a distinct "S" shape with a brighter bar of material across the center obscuring the central star. Although observable directly, it does show a moderate response to use of an OIII filter. It lies in a rich field of stars to the SE of m Centauri and has the common name "Spiral Planetary," despite some re-classifications. A nice object for most telescopes. (Image taken by Bexhill High School with Faulkes Telescope South operated by Las Cumbres Observatory Global Telescope Network. Used with permission.)

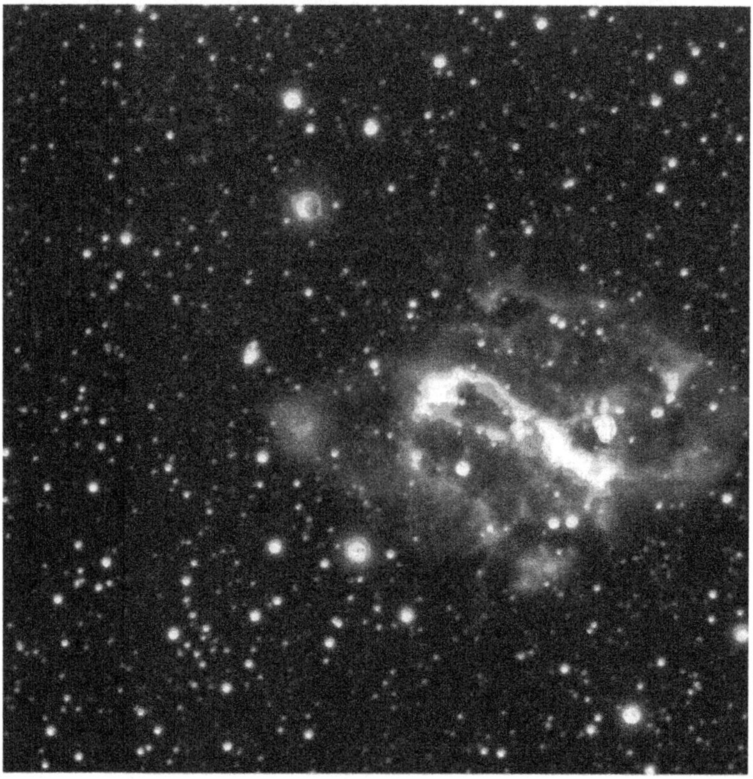

Fig. 8.66 NGC 5189

NGC 5194 (Messier 51) Canes Venatici

RA: 13h 29m
Dec: 47.11
Size: 6×5′
Magnitude: 8.4
Filters: Hα, NB, RGB

The incredible "Whirlpool Galaxy" in Canes Venatici, this is one of an interacting pair with NGC 5195, and the distorted arms are the clearest visual and photographic indication of this interaction. Lying 25 million light years away this galaxy is a favorite target for observers, and as it is a face-on spiral it makes a very pleasing object filled with details. HII and starburst regions abound across its disk and dust lanes and tails of stars thrown out by the interaction are visible even in relatively short exposures. Several supernovae have been recorded in this galaxy in recent years, so it is always worth keeping an eye on. (Image taken by JFS School with Faulkes Telescope North operated by Las Cumbres Observatory Global Telescope Network. Used with permission.)

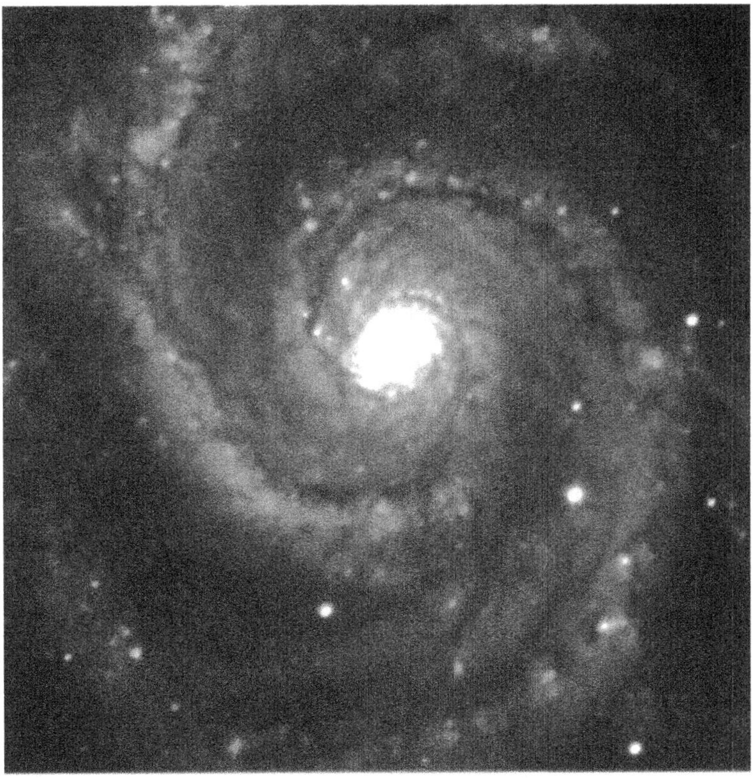

Fig. 8.67 NGC 5194

Messier 83 Hydra

RA: 13h 37m
Dec: −29.51
Size: 13 × 11′
Magnitude: 7.5
Filters: Hα, NB, RGB

The southern “Pinwheel Galaxy” is a beautiful barred spiral galaxy that is another favorite target for photographers, as it lays face-on to us, and the barred center is highly detailed. HII regions dot the arms as they whirl out into space, and the orientation gives the observer the advantage of seeing starburst regions, HII areas, dust lanes and the creamy central bar at the nucleus. The galaxy, which lies only 14 million light years away, is richly detailed and fairly large, spanning 12′ in total. (Image taken by Martin Griffiths with Faulkes Telescope South operated by Las Cumbres Observatory Global Telescope Network.)

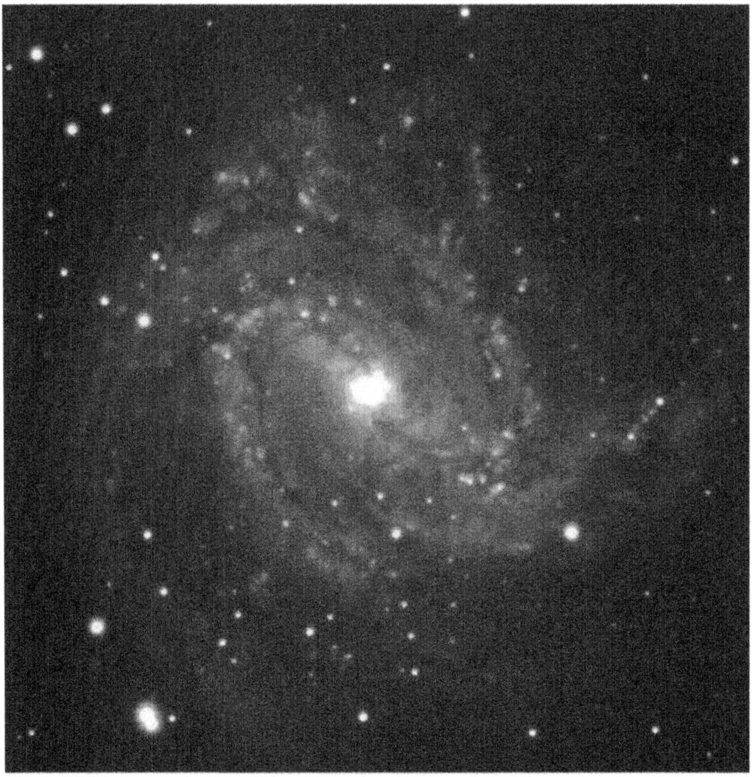

Fig. 8.68 M83

Messier 101 Ursa Major

RA: 14h 03m
Dec: −54.20
Size: 29 × 28′
Magnitude: 8.3
Filters: Hα, NB, RGB

The northern "Pinwheel Galaxy" is quite a contrast to its southern namesake! Fainter and more extended, with a distance of 21 million light years, M101 is still a good target for photographers, and the area surrounding it abounds in faint galaxies that can also be caught in the same field of view. The galaxy itself has extensive spiral structure and as it lies face on to us, many HII regions and blue knots of star formation are visible across the disk s are the myriad of dusty globules and lanes. In 2011 a bright supernova was recorded in this galaxy, the fourth such in a century. (Image taken by Robert Smith Academy with Faulkes Telescope North operated by Las Cumbres Observatory Global Telescope Network. Used with permission.)

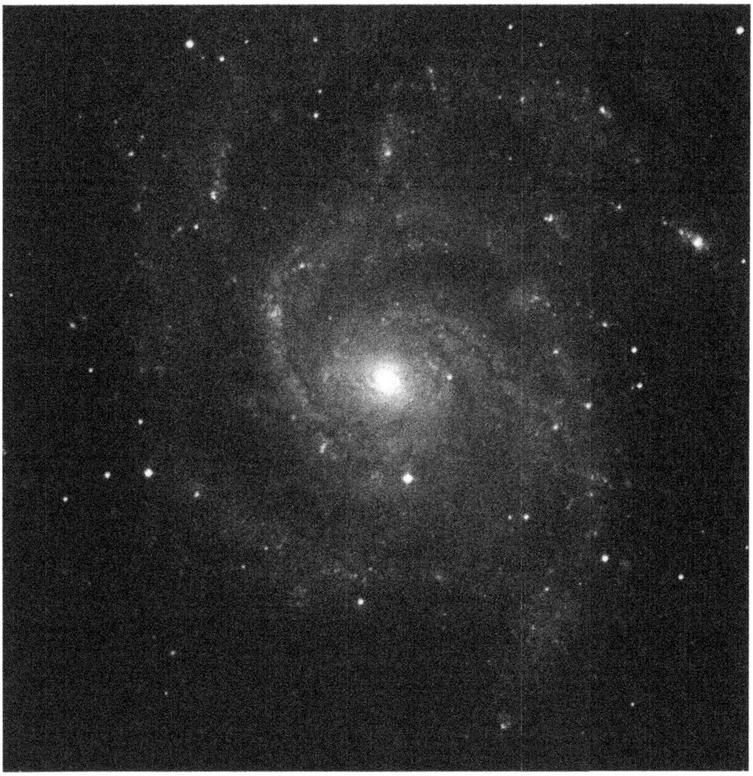

Fig. 8.69 M101

NGC 5907 Draco

RA: 15h 15m
Dec: +56.19
Size: 13×2′
Magnitude: 11.1
Filters: Hα, NB, RGB

This lovely edge-on spiral galaxy in Draco is thought to have vast numbers of dwarf stars as its major component. Photographically the dust lanes across the disc are clearly evident, as is the extensive nucleus fading into reddish arms that become faintly blue at their extremities. In 2006 an international group of astronomers discovered that this object has a large tidal stream of stars that loop around NGC 5907 twice and are possibly the result of some sort of collision with a dwarf galaxy that has no discernable nucleus. (Image taken by David Bowdley with Faulkes Telescope North operated by Las Cumbres Observatory Global Telescope Network. Used with permission.)

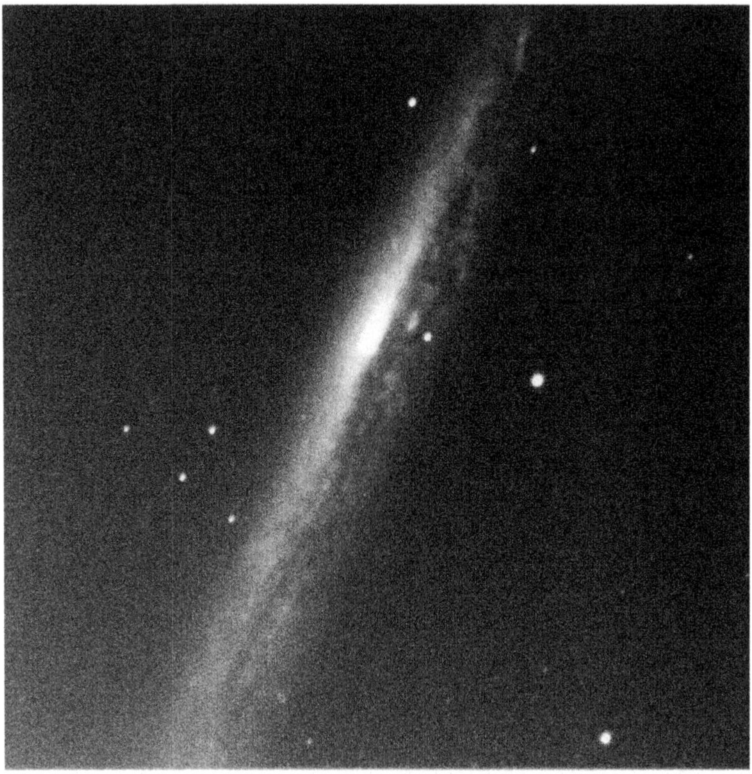

Fig. 8.70 NGC 5907

Messier 4 Scorpius

RA: 16h 23m
Dec: −26.31
Size: 26×25′
Magnitude: 5.8
Filter: RGB

A lovely large and highly resolvable globular cluster in Scorpius, M4 is visible in binoculars, and taking an image of the cluster is quite easy, revealing a strange little chain of stars across the nucleus. Lying at a distance of 7,200 light years this globular is a rather loose scattering with some concentration toward the center and made up of two distinct populations of stars—an uncommon detail in such objects, which are generally thought to be ancient and almost unchanged since the formation of the galaxy. This looseness has been attributed to M4 having passed through our galaxy many times and suffering tidal shock, which thins out the numbers of stars associated with the cluster. It is possible that this cluster was more massive and concentrated in the past. (Image taken by Bury College with Faulkes Telescope North operated by Las Cumbres Observatory Global Telescope Network. Used with permission.)

Fig. 8.71 M4 (*Credit*: Martin Griffiths)

IC 4604 Scorpius

RA: 16h 30m
Dec: −26.20
Size: 150×150′
Magnitude: 7.4
Filters: Hα, NB, RGB

This enormous ρ Ophiuchi complex surrounds the stars Antares in Scorpius and extends away into the background Milky Way for many degrees around. The large cloud making it up has components of reflection and HII nebulae and is suffused with dark nebulae that snake their way across the dim star fields of the galaxy here. The most obvious component is the long dust lane B144 leading directly to the complex above Antares, connecting B42 with the globular cluster M4 visible at the bottom of the cloud but not associated with it. This large object will be a great target for a photo mosaic or simply for a digital SLR with a telephoto lens, as long as the mount is driven. A wonderful area to explore and record. (Image taken by Martin Griffiths at Griffon Educational Observatory, Andalucia Spain.)

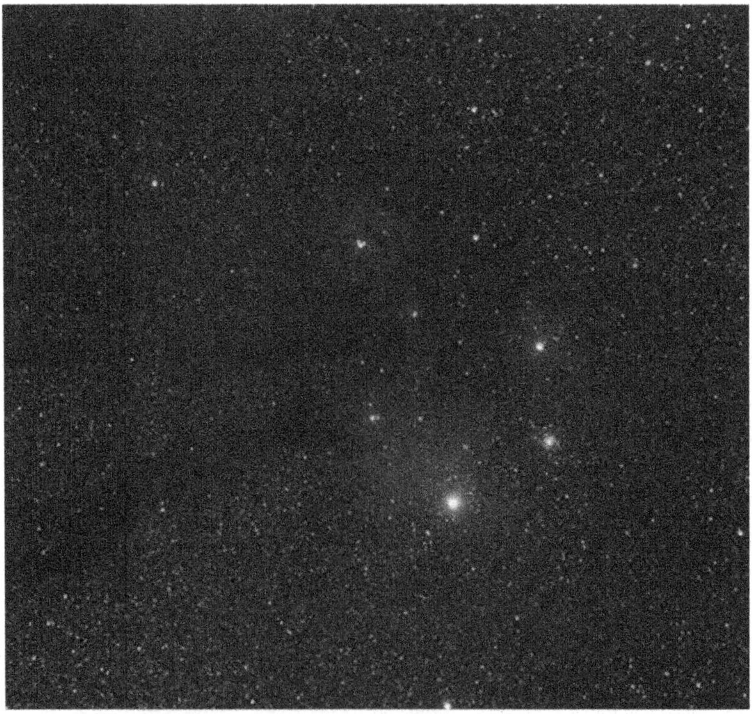

Fig. 8.72 IC 4604

NGC 6188 Ara

RA: 16h 40m
Dec: −48.47
Size: 20 × 12
Magnitude: 8.5
Filters: Hα, NB, RGB

A large nebulous area in the constellation of Altar, with the open cluster NGC 6193 as the major visual draw at magnitude 5.2. Of immediate interest to any observer, the nebulosity surrounding the cluster is fairly faint and needs long exposure to bring out the fill glory of its color, dust lanes and ionization fronts. Lying at 4,000 light years away the nebula, nicknamed the "Fighting Dragons," spreads over a square degree from the star cluster of which the brightest two stars are responsible for the ionization of the gas cloud. (Image by Heather Patton, Faulkes telescope south. Image courtesty of LCOGT.)

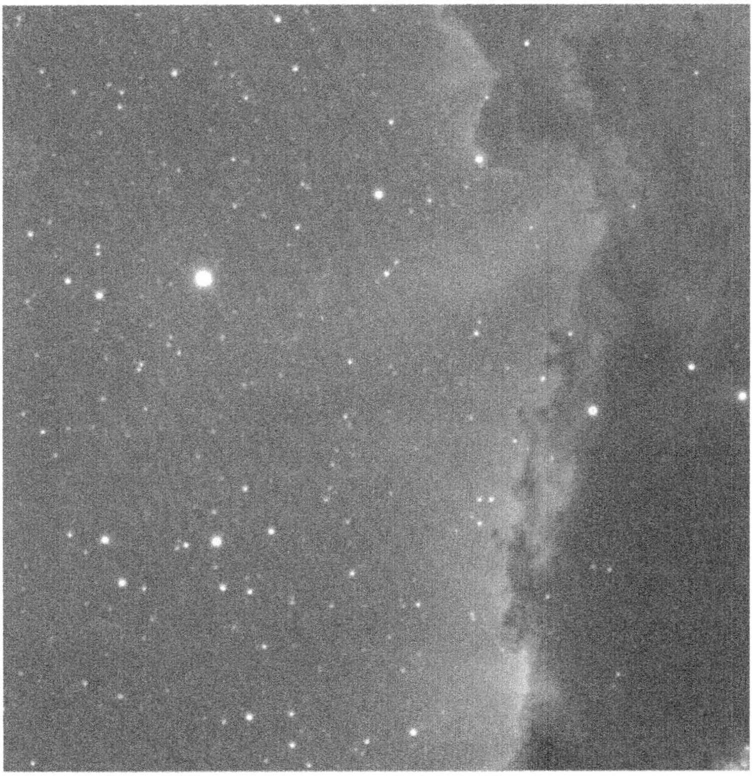

Fig. 8.73 NGC 6188

NGC 6302 Scorpius

RA: 17h 13m
Dec: −37.06.2
Size: 83″
Magnitude: 9.6
Filters: Hα, OIII, SII, RGB

The fantastic "Bug Nebula" is a beautiful complex of gray-blue gas extended on an E–W axis and shining at ninth mag. The planetary nebula has obvious extensions that can be seen directly in a low power that extend the object to over 1′ across. Careful observation reveals the W side to be the brighter part after the central condensation, which looks star-like at first glance but is in fact a block of white light, with no starry object within. NGC 6302 is a wonderful detailed object that higher powers reveal much of, and it is also cataloged as Caldwell 69. It is a showpiece object that can be viewed in most instruments, although observers in high northern latitudes will be unable to see it, as it will lie below the horizon. (Image taken by Martin Griffiths with Faulkes Telescope North operated by Las Cumbres Observatory Global Telescope Network.)

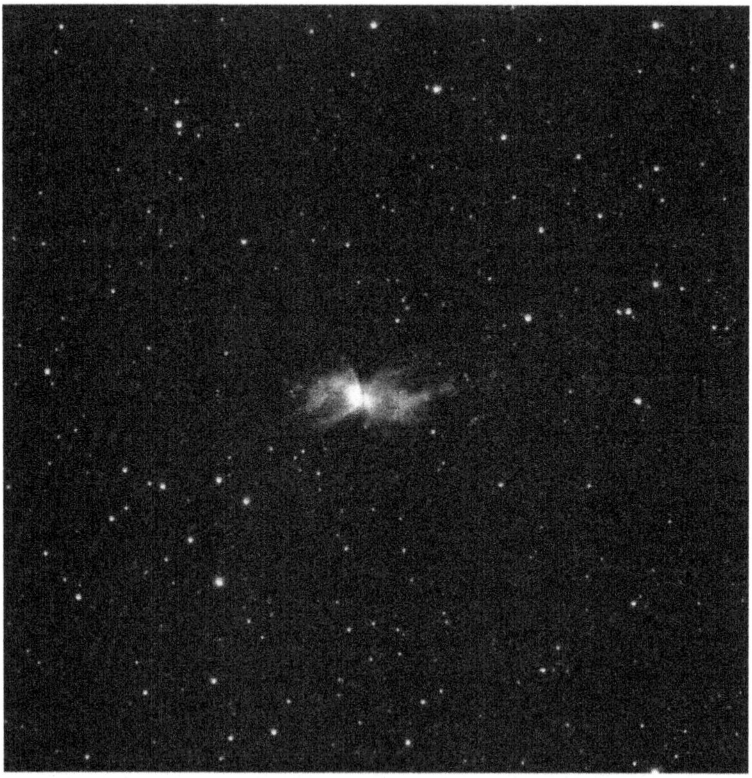

Fig. 8.74 NGC 6302

NGC 6337 Scorpius

RA: 17h 22m
Dec: −38.29.00
Size: 50″
Magnitude: 12.3
Filters: Hα, OIII, SII, RGB

NGC 6337 is invisible from high northern latitudes, as it lies just 2.5° west from Shaula in the tail of the Scorpion and takes some finding in this rich field of stars, as the shape is indistinct in a low power field. Use of an OIII filter is helpful and delivers a lovely ring-like object that is obvious but faint. Higher magnifications bring out this ring feature more clearly and allow the observer to see why this object is known as the "Cheerio Nebula" in the United States. Due to its area and shape, it seems a little fainter than its quoted magnitude. Ghostly. (Image taken by Martin Griffiths with Faulkes Telescope North operated by Las Cumbres Observatory Global Telescope Network.)

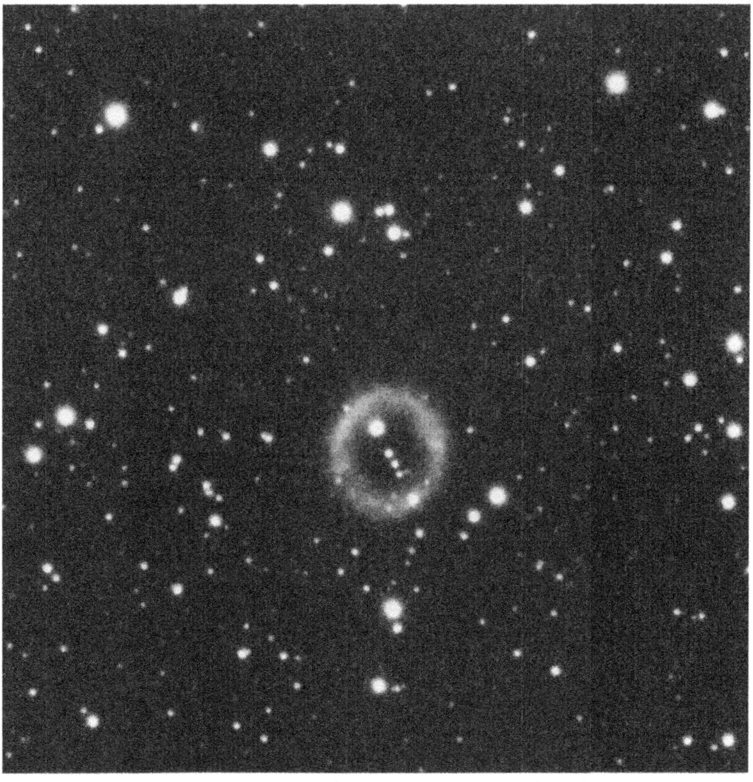

Fig. 8.75 NGC 6337

Messier 20 Sagittarius

RA: 18h 02m
Dec: −23.01
Size: 30×28′
Magnitude: 6.1
Filters: Hα, NB, RGB

The "Trifid Nebula" is a common astrophotography target and can be seen by most observers in the northern and southern hemispheres. This wonderful cloud of ionized gas is broken into three parts by dark dust lanes and radiates in Hα, with a northern section that shines with the blue glow of a reflection nebula. Lying at 5,200 light years away this bright nebula is lit by a small cluster of extremely massive stars, some of which are still emerging from their birth shrouds in the gas. The nebula has been intensively studied by the Hubble Space Telescope, revealing many areas where EGGS (evaporating gaseous globules) are present, hinting at a further burst of star formation in this mesmerizing object. (Image by Martin Griffiths taken at Griffon Educational Observatory, Andalusia, Spain.)

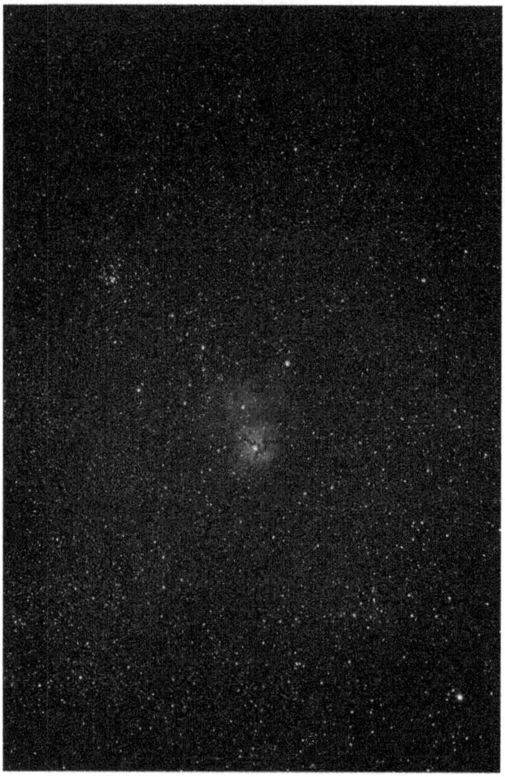

Fig. 8.76 M20

Messier 8 Sagittarius

RA: 18h 03m
Dec: −24.23
Size: 55×40′
Magnitude: 4.9
Filters: Hα, NB, RGB

Lying to the south of M20 is the larger, more complex structure of the "Lagoon Nebula." In actuality the presence of both nebulae in this area is merely a line of sight effect, as M8 lies 4,100 light years away and is resplendent in stars and glowing with Hα light. Many dark "Bok" globules can be seen against the red gaseous background—testament to ongoing star formation within the cloud. It is an easy target for photographers, and small field CCD's will have little trouble in making a mosaic of this large cloud—over three times the diameter of the Moon and studded with bright stars leading to an off-center concentration of light where a new cluster is beginning to peep through. (Image taken by Martin Griffiths, Faulkes telescope south. Image courtesty of LCOGT.)

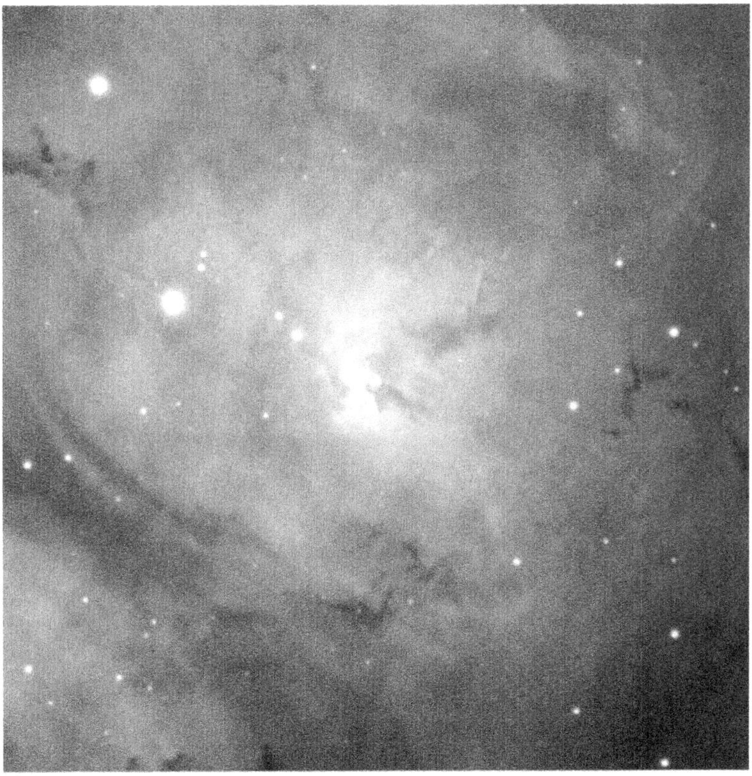

Fig. 8.77 M8

NGC 6572 Ophiuchus

RA: 18h 12m
Dec: −06.51.2
Size: 16″
Magnitude: 8.1
Filters: Hα, OIII, SII, RGB

Although this planetary has the common name "Blue Racquetball," it is also known as the "Emerald" or "Turquoise" orb! It seems disparate observers see it slightly differently. NGC 6572 is obvious at first glance in the field south of 71 Ophiuchi. Observing it with a low power reveals it as a wonderful light emerald green in color to some, this author included. In higher powers the nebula is extended across a N/S axis, and the interior seems to separate a little from its outer edges. A lovely object, as the color makes up for its small size! It always seems a little larger than the quoted dimensions of 16″—it may actually be over 25″ across. (Image taken by Ed Gomez with Faulkes Telescope North operated by Las Cumbres Observatory Global Telescope Network. Used with permission.)

Fig. 8.78 NGC 6572

Messier 16 Serpens

RA: 18h 18m
Dec: −13.49
Size: 40×35′
Magnitude: 5.8
Filters: Hα, NB, RGB

The huge and beautiful "Eagle Nebula" in Serpens is another favorite target of observers and photographers alike. This HII region has several dark nebulae associated with it—including the famed "Pillars of Creation" at the center of the object. The distance to M16 is 7,000 light years, and over 450 stars have been counted within the nebula, most of which are hot O- or B-type stars, about 5 million years old at most, that are responsible for illuminating various portions of the cloud. CCD imaging should bring out the pillars and the "Stellar Spire" area to the east of the pillars. The area glows with hydrogen gas and is studded with bright stars. The whole complex is unforgettable. (Image taken by Martin Griffiths.)

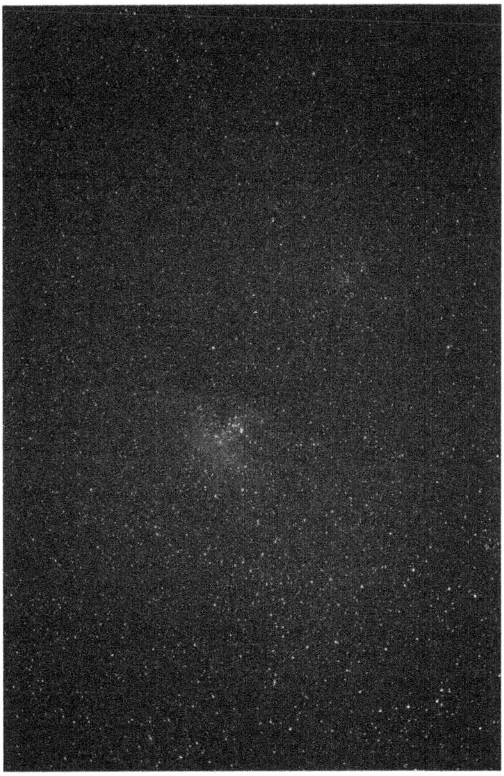

Fig. 8.79 M16

Messier 17 Sagittarius

RA: 18h 20m
Dec: −16.01
Size: 25×15′
Magnitude: 5.8
Filters: Hα, NB, RGB

The "Omega" or "Swan Nebula" (M17) is a huge complex of newborn stars still shrouded in hydrogen gas, making the cloud glow brightly and more luminous than M16 to its north. At 5,500 light years away this complex is quite large and contains 30,000 solar masses of material. It has been observed for many years by astronomers, and the large bar-like 'body" of the swan is easily discernable, as is the neck and head, albeit that the complex is seen upside down from northern climes. To bring out the outer regions of swirling gas requires an Hα filter and slightly longer exposures, but the tendrils of material are worth the additional effort. (Image taken by Martin Griffiths at Griffon Educational Observatory, Andalucia, Spain.)

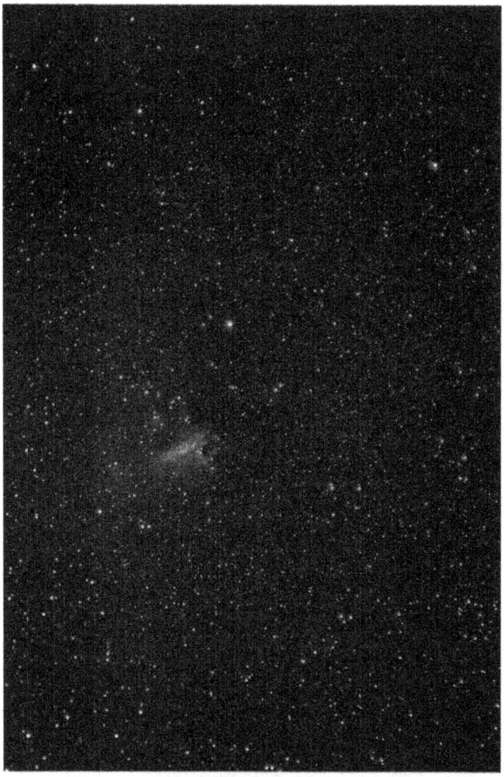

Fig. 8.80 M17

Messier 57 Lyra

RA: 18h 53m
Dec: +33.01.7
Size: 86″
Magnitude: 8.8
Filters: Hα, OIII, SII, RGB

This is the superb "Ring Nebula," M57, in Lyra. Even in small apertures the ring shape is immediately evident, as is the slightly purple hue to it that tends to red at the outer edges in higher magnifications, with a well-defined dark, crisp center but no hint of the central star in small apertures. In larger telescopes the color is more greenish and the proportions of the ring are clearer and revealed as slightly oblate E/W under careful scrutiny, with a filled, veiled center through which is visible the 14th magnitude central star. Fabulous! (Image taken by Martin Griffiths with Faulkes Telescope North operated by Las Cumbres Observatory Global Telescope Network.)

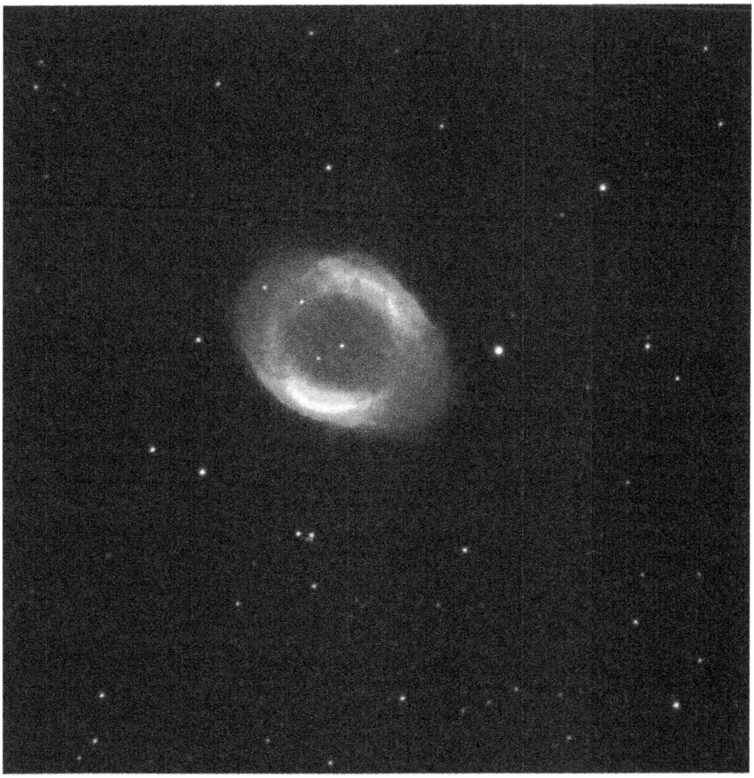

Fig. 8.81 M57

NGC 6781 Aquila

RA: 19h 18m
Dec: −06.32.2
Size: 120″
Magnitude: 11.4
Filters: Hα, OIII, SII, RGB

NGC 6781 is a beautiful, ghostly nebula in the heart of Aquila. Best to star hop from δ Aquilae, where after locating the field the observer should be rewarded with a large blue-gray patch of light almost 2′ across, which is distinctly fainter on the southern portion even at low powers. The nebula is revealed as an 11th magnitude halo with a thicker N edge and a central darkening that is irregular and mottled yet still opaque to the central star. It can be seen in apertures less than 150 mm under good conditions and is also a Herschel 400 catalog object. (Image taken by Martin Griffiths with Faulkes Telescope North operated by Las Cumbres Observatory Global Telescope Network.)

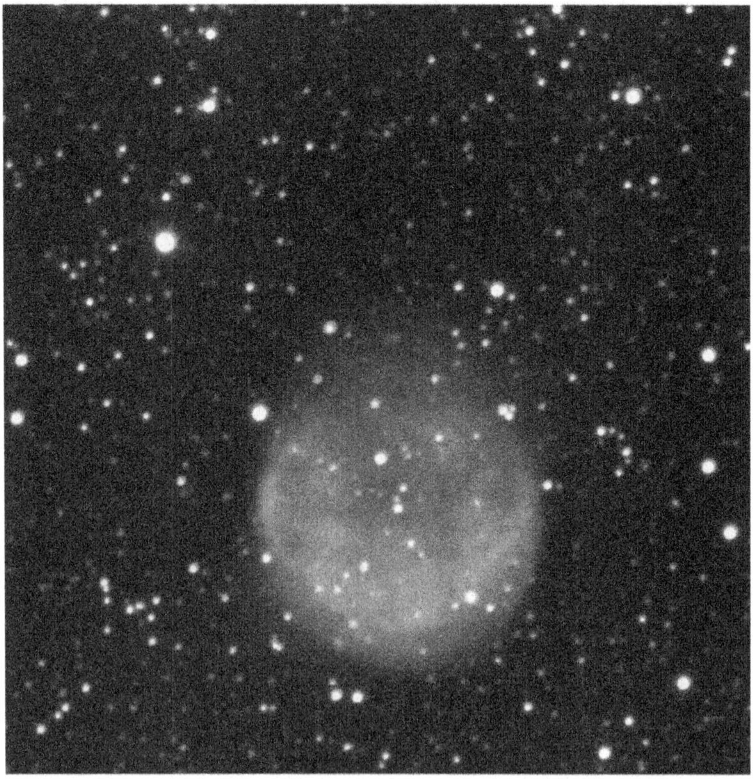

Fig. 8.82 NGC 6781

NGC 6818 Sagittarius

RA: 19h 43m
Dec: −14.09.2
Size: 22″
Magnitude: 9.3
Filters: Hα, OIII, SII, RGB

The "little gem" planetary just to the N of the large NGC 6822 ("Barnard's Galaxy") is much easier to see than this extended and faint object! It is revealed as a glowing turquoise object with a slightly elliptical shape, elongated N/S and about 25″ across and plainly evident due to brightness and color even at low power. In moderate powers it looks evenly filled, though averted vision may show a slight central darkening, but no central star visible. NGC 6818 displays a vivid color and can be seen in most backyard telescopes. It is a Herschel 400 object, too. (Image taken by Martin Griffiths with Faulkes Telescope North operated by Las Cumbres Observatory Global Telescope Network.)

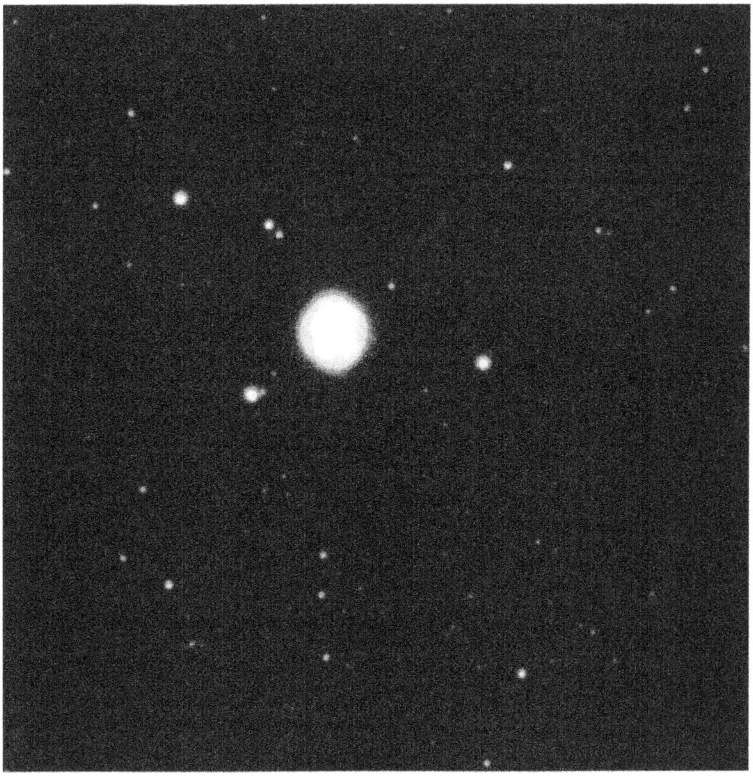

Fig. 8.83 NGC 6818

NGC 6822 Sagittarius

RA: 18h 44m
Dec: −14.47
Size: 15×13′
Magnitude: 9.3
Filters: Hα, RGB

This is the tiny and difficult "Barnard's Galaxy" in Sagittarius, a local group dwarf galaxy that has the distinction of being separated into stars so that Edwin Hubble used it as a distance indicator to prove the extragalactic nature of these objects. Generally ovoid in shape, it has an apparent magnitude of 9.3, but this is spread over an area half the size of the full Moon, rendering the galaxy almost invisible with the spread of light. It has many small HII regions that may be brought out with careful editing. It is the historical context of this object and the type of galaxy that Hubble later defined coupled to its difficulty to image that make it a challenge for the observer. (Image by Martin Griffiths taken at Griffon Educational Observatory, Andalucia, Spain.)

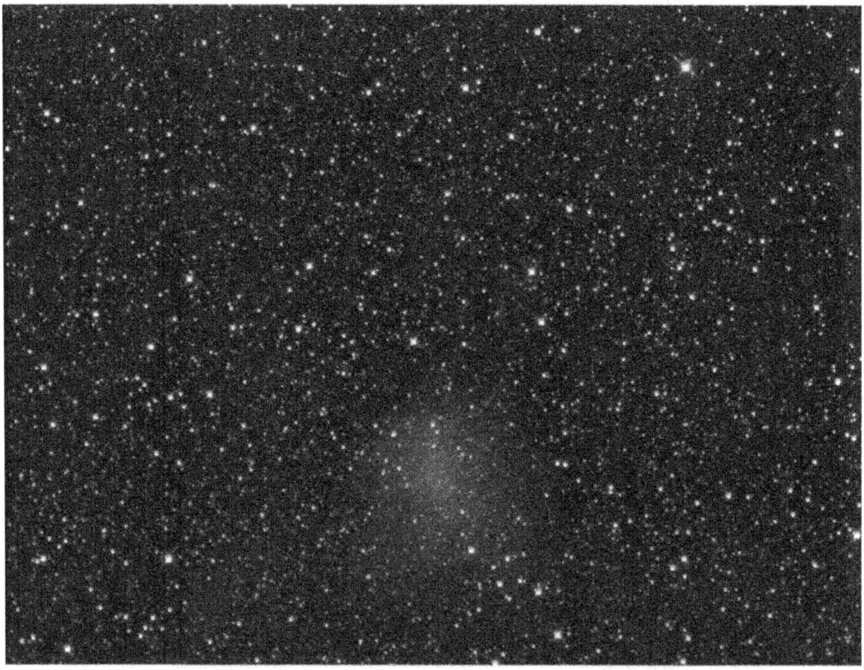

Fig. 8.84 NGC 6822

NGC 6826 Cygnus

RA: 19h 44m
Dec: +50.31.5
Size: 30″
Magnitude: 8.8
Filters: Hα, OIII, SII, RGB

This is the fabulous "Blinking Nebula" in NE Cygnus discovered by William Herschel. In low power the turquoise color is beautiful and contrasts wonderfully with the yellow binary 16 Cygni in the same low power field of view. In higher powers the nebula is a little elongated along a N/S axis, and there is a central bright concentration but no glimpse of the central star. Under high magnifications the outer part of the nebula does have a slight "atmosphere," or edge, even in direct vision. NGC 6826 is a spectacular object with a vivid color that should be visible in telescopes of all apertures. (Image taken by Martin Griffiths with Faulkes Telescope North operated by Las Cumbres Observatory Global Telescope Network.)

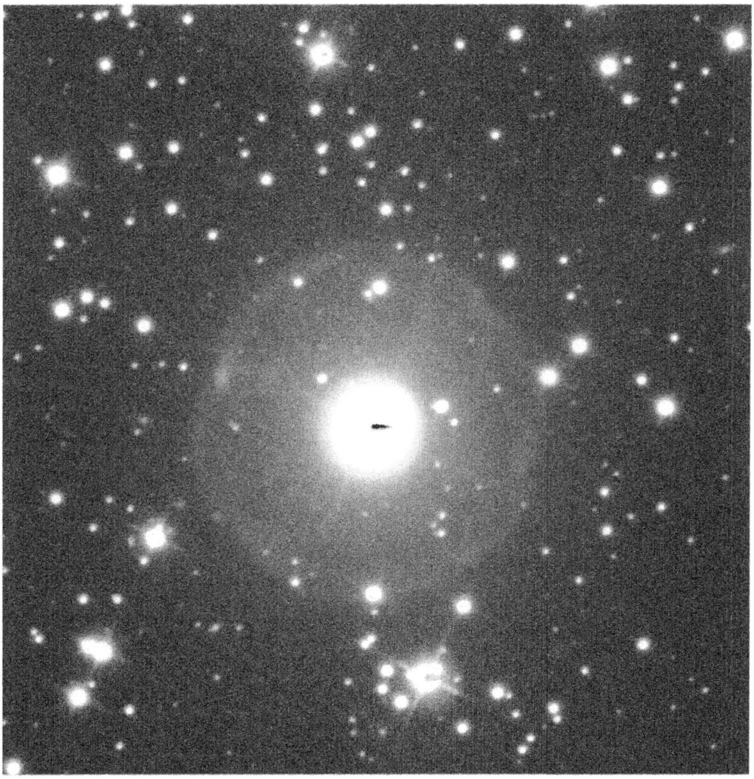

Fig. 8.85 NGC 6826

Messier 27 Vulpecula

RA: 19h 59m
Dec: +22.43.3
Size: 8′
Magnitude: 7.3
Filters: Hα, OIII, SII, RGB

NGC 6853 is otherwise known as the "Dumb Bell Nebula," M27, in Vulpecula. The double V shape of this wonderful planetary is evident at low powers, and the whole shell is blue-gray and fairly bright. The center looks ghostly, yet detailed no matter what instrument or power. In most telescopes M27 is an amorphous blob of light with thicker edges to the E/W. It has a great central bar and a faint central star coupled to a knotty appearance across the disk throughout the thinner sections to N/S of the lobes. It is also filled with at least ten countable stars and is a beautifully detailed object in any telescope. (Image taken by Martin Griffiths with Faulkes Telescope North operated by Las Cumbres Observatory Global Telescope Network.)

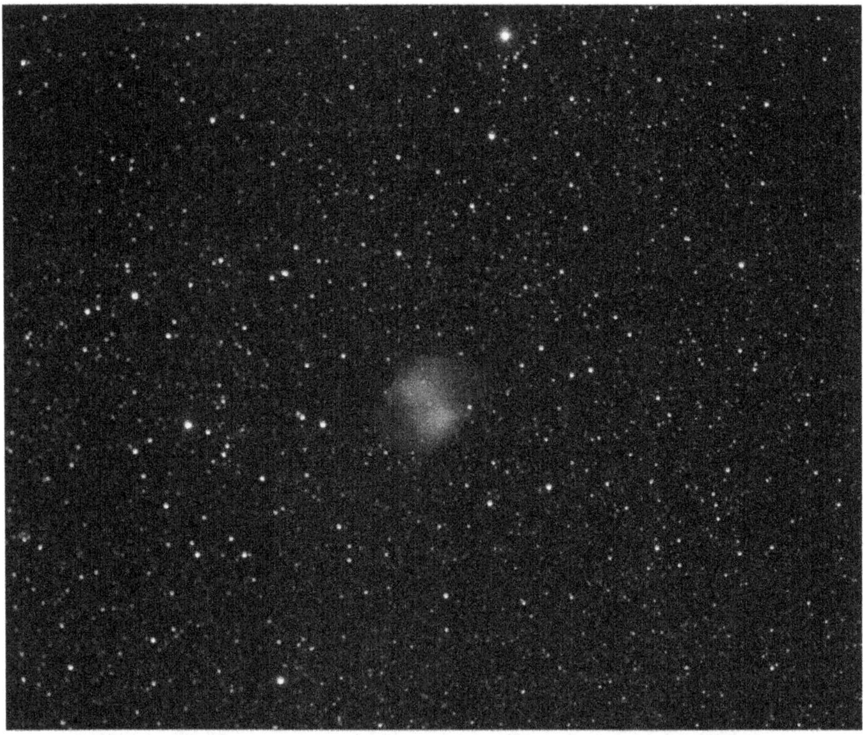

Fig. 8.86 M27

NGC 6888 Cygnus

RA: 20h 12m
Dec:+38.21
Size: 418 × 12′
Magnitude: 7.6
Filters: Hα, NB, RGB

The "Crescent Nebula" in Cygnus, also known as Caldwell 27, is actually a cloud of battered interstellar medium and fast gas outflow from a Wolf-Rayet star. The nebula is broadly crescent-shaped, but in longer exposures it actually has a lot of structure in the form of sheets and tendrils of gas that create an oval nebula with the bright W-R star embedded within. Visually a nice object, CCD imaging should bring out all these details. This Hα image should reveal the filamentary structure of the nebula near the W-R star. (Image taken by Martin Griffiths with Faulkes Telescope North operated by Las Cumbres Observatory Global Telescope Network.)

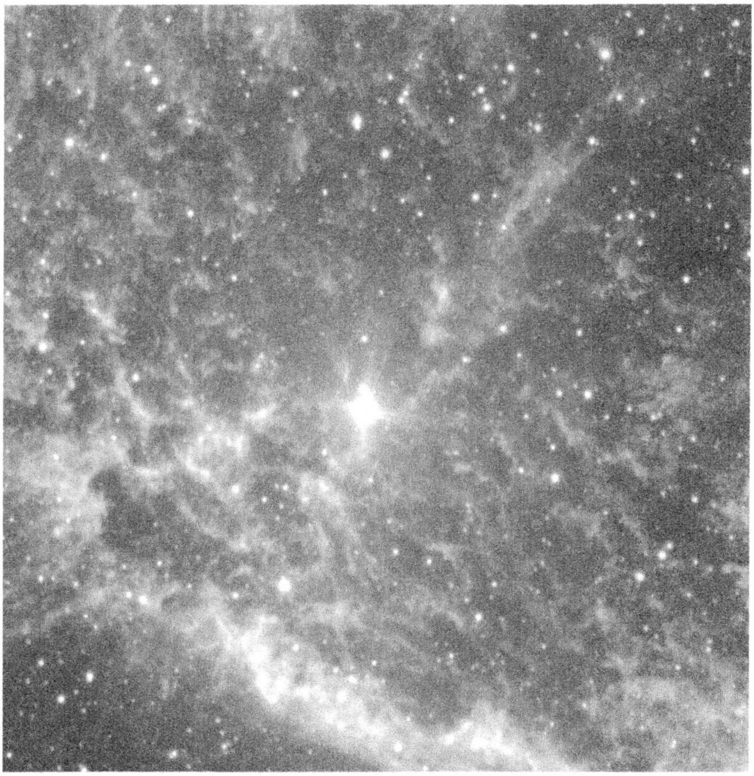

Fig. 8.87 NGC 6888

NGC 6905 Delphinus

RA: 20h 22m
Dec: +20.06.3
Size: 42″
Magnitude: 11.1
Filters: Hα, OIII, SII, RGB

NGC 6905 is a lovely little planetary in Delphinus enjoying a beautiful Milky Way background that makes it a little challenging to star hop to but is worth the effort, as this planetary is intensely blue and obvious in the field. A moderate power brings out a slight mottling to its interior and shows the elongation on a N/S axis a little better. With averted vision the center gets brighter, though in direct vision it looks darker! A wonderful deep sky object that is also known as the "Blue Flash." NGC 6905 is a Herschel 400 object, too. (Image taken by Martin Griffiths with Faulkes Telescope North operated by Las Cumbres Observatory Global Telescope Network.)

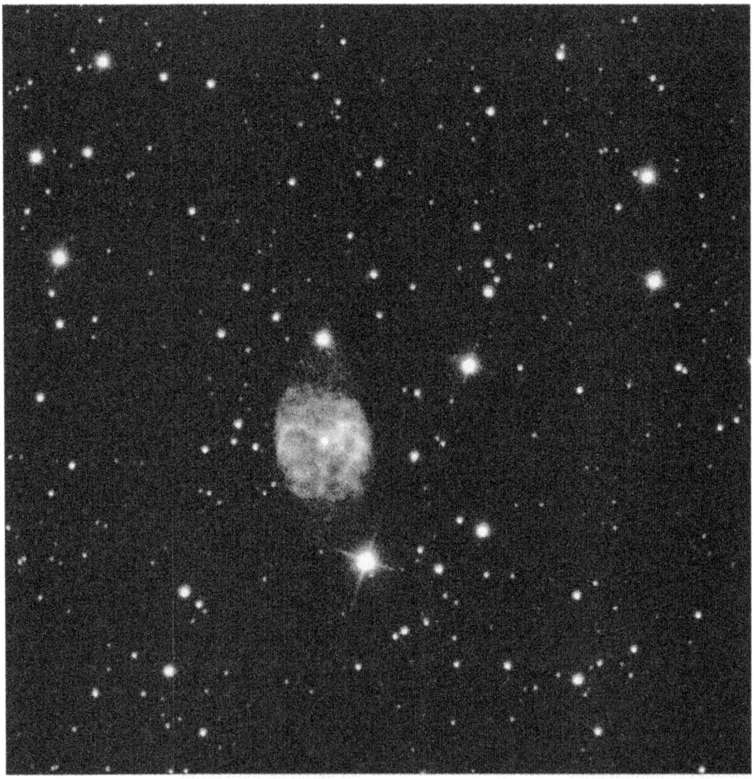

Fig. 8.88 NGC 6905

NGC 6960 Cygnus

RA: 20h 45m
Dec: +30.42
Size: 180×25′
Magnitude: 7.1
Filters: Hα, SI, OIII, NB, RGB

This is the famous "Witches Broom Nebula" that is actually a part of the vaster Veil Nebula complex. NGC 6960 is centered on the star 52 Cygni and sweeps past it to spread out like a bundle of twigs—the broom of its name. It shows a lovely striated pattern of gas with smaller feathery extensions in long-exposure photographs. It is the remnants of a supernova that exploded over 8,000 years ago. The distance to this feature is reported to be about 1,450 light years, and the area glows with the light of oxygen sulfur and hydrogen, so use of many of the above filters is recommended to bring out the best in this lacy, delicate web of gas expanding across the Milky Way. (Image taken by Martin Griffiths at the Griffon Educational Observatory, Andalucia, Spain.)

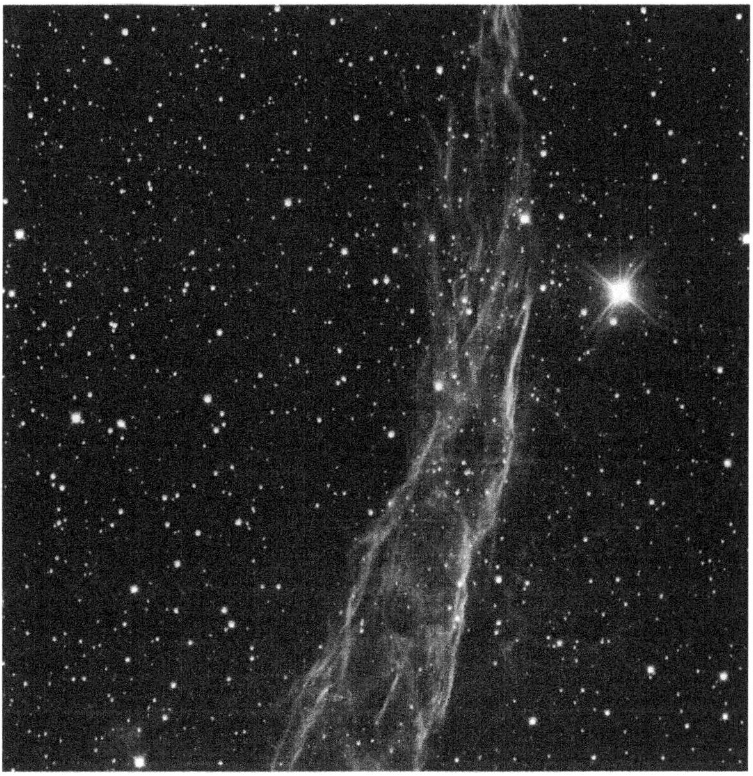

Fig. 8.89 NGC 6960

NGC 6992 Cygnus

RA: 20h 45m
Dec: +30.42
Size: 180×25′
Magnitude: 7.1
Filters: Hα, NB, RGB

The eastern part of the "Veil Nebula," NGC 6992 is larger than the Witch's Broom Nebula, and with a large field of view the photographer may capture the entire "Cygnus Loop," including Pickering's triangle, too. Visually this looks a bit like an arcing rainbow, with subtle colors of red and blue overall, but use of an OIII filter also brings out some detail. Photographically the nebulae is stunning—lots of wispy detail extending over an area more than 2° across. (Image taken by Martin Griffiths.)

Fig. 8.90 NGC 6992

IC 5067 Cygnus

RA: 20h 52m
Dec: +44.25
Size: 50×60′
Magnitude: 4.3
Filters: Hα, NB, RGB

The "Pelican Nebula," close to the fabulous NGC 7000, is a fainter object to capture, and with its large field it may be possible to mosaic the images if the observer cannot contain it in the field of the CCD. It gets its name from its resemblance to the large beaked bird, though in reality it looks more like a pterodactyl than a pelican. Lying at 1,800 light years away this cloud has an actual extent of 60 light years or more and is crossed by many dark lanes that outline the shape of its avian namesake. (H-alpha image of the Pelican nebula at center-right taken by Andy Burns of the Wiltshire Astronomical Society. Used with permission.)

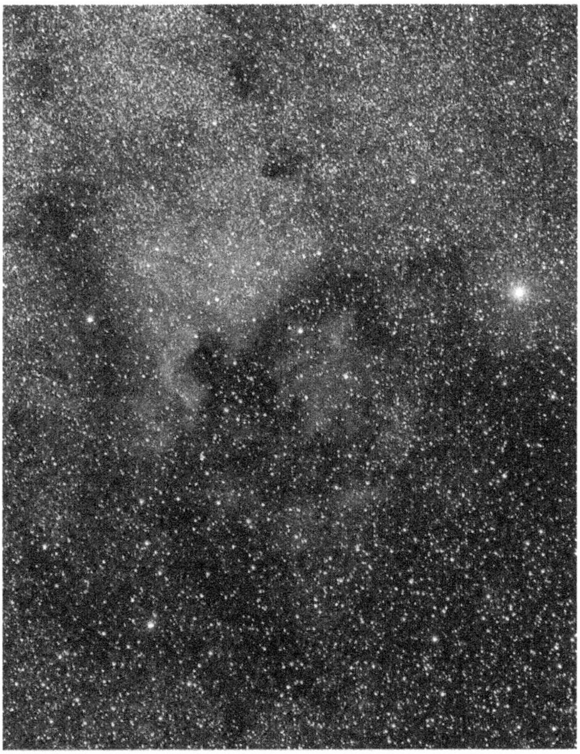

Fig. 8.91 IC 5067

NGC 7000 Cygnus

RA: 20h 59m
Dec: +44.31
Size: 120×100′
Magnitude: 4.3
Filters: Hα, NB, RGB

The wonderful "North American Nebula" in northern Cygnus is huge cloud of gas visible to the naked eye from a dark sky site, and yet it is one of the more difficult objects to image, due to its size and extent and relatively low luminosity despite an integrated magnitude of 4. At four times the size of the full Moon a wide field is required to capture it all at once. It is at the same relative distance as the Pelican Nebula, and it is not known which star is producing the ionization of the nebula. It has been suggested that Deneb (α Cygni) is responsible for this illumination, but no one is quite sure of this—both star and nebula lie at approximately the same distance, but this is not definitely confirmed. (Image taken by Martin Griffiths at the Griffon Educational Observatory, Andalucia, Spain.)

Fig. 8.92 NGC 7000

NGC 7008 Cygnus

RA: 21h 00m
Dec: +54.32.6
Size 100″
Magnitude: 10.7
Filters: Hα, OIII, SII, RGB

This fantastic nebula in northern Cygnus reveals itself as a white, smallish hook-shaped object on first appearance in low powers and lies close to a ninth mag star on its S edge. Moderate to high magnifications bring out finer detail connecting two lobes together and rendering the object a little rounder. However, in most 'scopes the object remains a rather disappointing gray-white in color. NGC 7008 is amenable to direct vision though it does respond quite well to use of an OIII filter. It is also a Herschel 400 object that is visible through modest apertures. (Image taken by Oliver Tunnah with Faulkes Telescope North operated by Las Cumbres Observatory Global Telescope Network. Used with permission.)

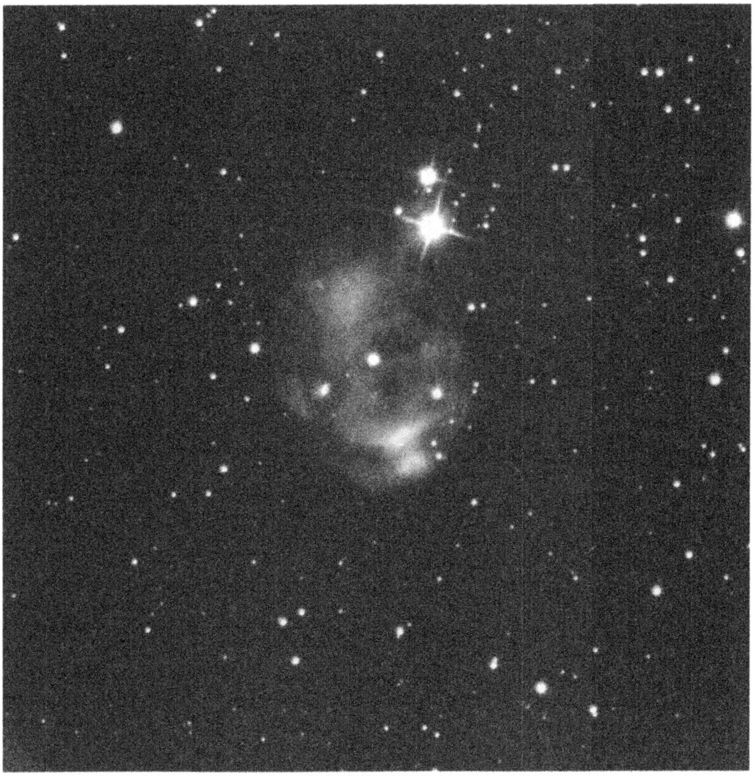

Fig. 8.93 NGC 7008

IC 5146 Cygnus

RA: 21h 53m
Dec: +47.16
Size: 15 × 12′
Magnitude: 12.1
Filters: Hα, NB, RGB

The lovely "Cocoon Nebula" in northern Cygnus is a favorite target for observers and photographers alike, as its position on the Milky Way fills the field with stars and the long dark dust lane, B 168, leading to this flowery-looking nebula is a good feature to look out for. IC 5146 is over 3,000 light years away and is just 12′ across but makes a lovely concentrated target for CCD imaging or even wide-field photography. Glowing with hydrogen light an Hα filter will bring out the best of this object. (Image taken by Martin Griffiths at the Griffon Educational Observatory, Andalucia, Spain.)

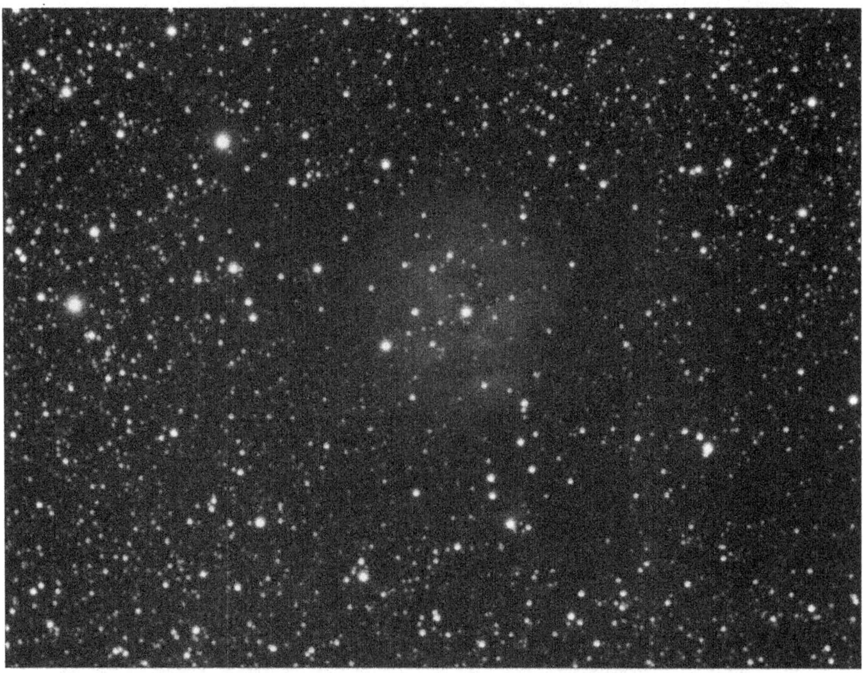

Fig. 8.94 IC 5146

NGC 7009 Aquarius

RA: 21h 04m
Dec: −11.21.8
Size: 44″
Magnitude: 7.8
Filters: Hα, OIII, SII, RGB

The stunning "Saturn Nebula" in Aquarius is visible in 'scopes of all apertures and reveals great detail in its elongated body and rays at higher powers. In low power these tiny "ansae," or FLIERS, plus the lovely green hue of the planetary are instantly visible, while moderate power does show some subtle patterning of the body of the nebula. In averted vision at higher powers, observers equipped with 'scopes of 20+cm will see small blobs on the end of the ansae. A fantastic object, and it comes as no surprise to learn that it is both Caldwell 55 and a Herschel 400 catalog object, too. (Image taken by Martin Griffiths with Faulkes Telescope North operated by Las Cumbres Observatory Global Telescope Network.)

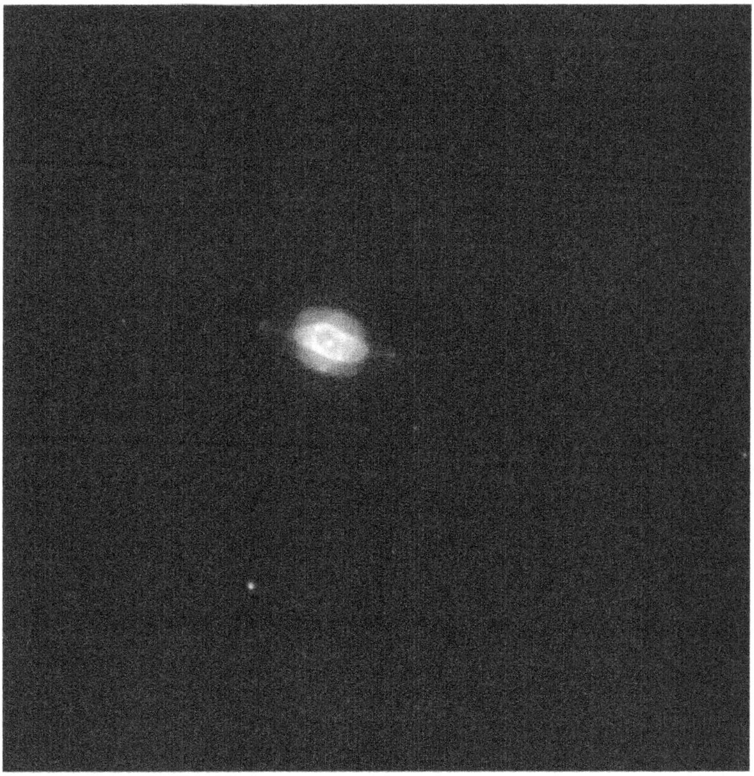

Fig. 8.95 NGC 7009

NGC 7023 Cepheus

RA: 21h 01m
Dec: +68.10
Size: 20×18′
Magnitude: 6.8
Filters: Hα, NB, RGB

The "Iris Nebula" in Cepheus is a bright reflection nebula that shows a distinct blue tinge even in short-exposure photographs and is surrounded by the darkness of dust clouds, with lanes intersecting at the nebula center. Lying 1,300 light years away the cloud seems to be centered on a small star cluster that is the NGC 7009, while the cloud itself has the designation LBN 487 ("Lynd's Bright Nebula") and is also a Caldwell object. The entire area is rich in pre-biotic molecules that are the building blocks of possible future life, so it's worth taking some time to capture and contemplate this lovely cloud in the depths of space. (Image taken by Martin Griffiths.)

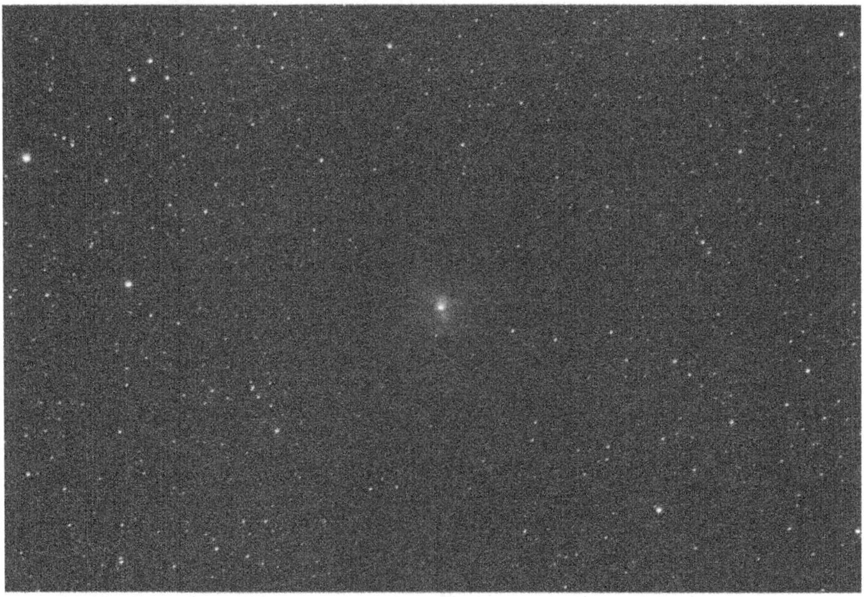

Fig. 8.96 NGC 7023

NGC 7293 Aquarius

RA: 22h 29m
Dec: −20.50.2
Size: 12′
Magnitude: 7.3
Filters: Hα, OIII, SII, RGB

The wonderful "Helix Nebula" in Aquarius is a glorious sight in telescopes of large aperture. But despite its magnitude, it remains nigh on invisible in smaller telescopes for users in high northern latitudes. It can be seen as a ghostly ring of green-gray light at low powers and is surprisingly large but shows a ring structure and a darker inner core quite easily. NGC 7293 looks best in 20×80 binoculars, as the field is larger and the combined light from the object becomes easier to grasp as it is better concentrated. The observer is then looking at it, not through it! NGC 7293 is a beautiful and fairly detailed object, which also has the catalog number Caldwell 63. (Image taken by Martin Griffiths at the Griffon Educational Observatory, Andalucia, Spain.)

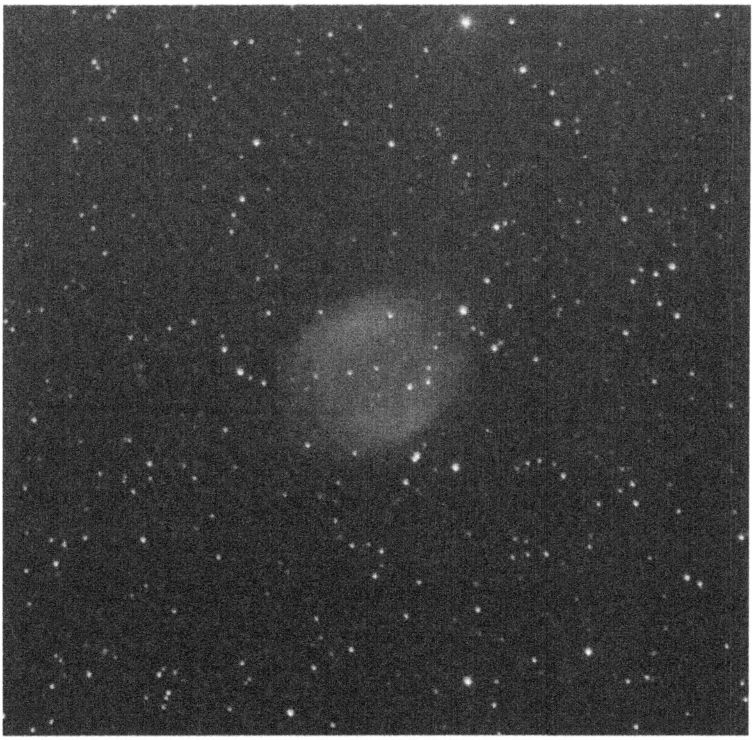

Fig. 8.97 NGC 7293

NGC 7635 Cassiopeia

RA: 23h 20m
Dec: +61.12
Size: 15 × 8′
Magnitude: 10.2
Filters: Hα, NB, RGB

The "Bubble Nebula" in Cassiopeia is a great target, as its bright, bubble shaped and part of an extended nebulous systems that reaches toward the star cluster M52. It is an extremely distant object for such a nebula, lying at a distance of 11,000 light years away but is visible at this distance as the shell itself is being illuminated by an energetic Wolf-Rayet star. In long exposures the whole shell of the bubble is evident, whereas short exposures will bring out a crescent shape around the star. The background nebulosity is open to capture in R band and Hα, too. (Image taken by Chris O'Morain with Faulkes Telescope North operated by Las Cumbres Observatory Global Telescope Network. Used with permission.)

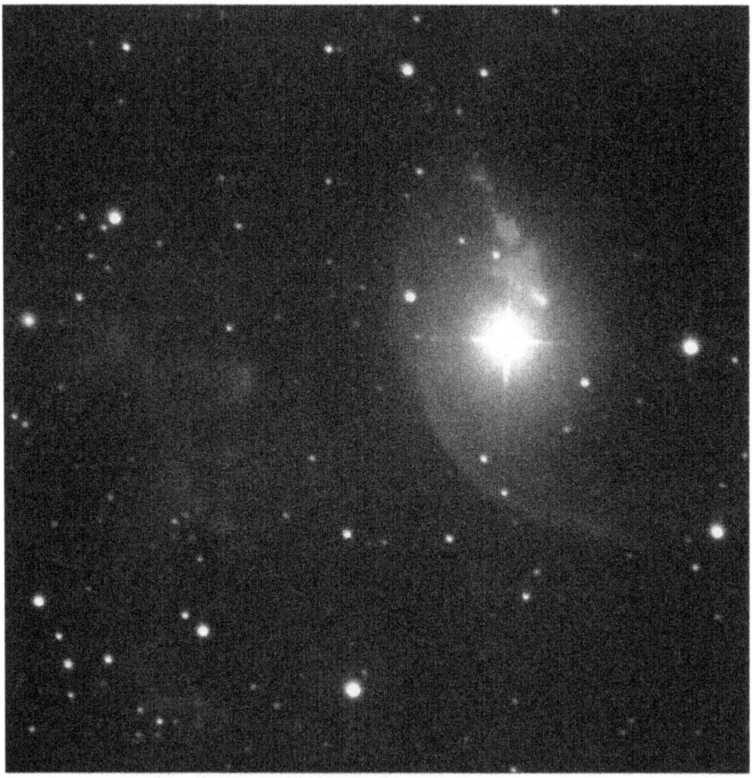

Fig. 8.98 NGC 7635

NGC 7662 Andromeda

RA: 23h 25m
Dec: +42.32.1
Size: 32″
Magnitude: 8.3
Filters: Hα, OIII, SII, RGB

The beautiful "Blue Snowball" planetary nebula in northern Andromeda is a wonderful object, as its intense blue color and mottled appearance in low powers renders it precisely as its name implies. It can be found SE of ι Andromedae and is an immediate, arresting sight as a blue ball over 30″ across standing out in the field of view. There is a central condensation visible in moderate powers that occasionally looks star-like. NGC 7662 is visible in any sized telescope, and low to moderate magnifications seem best to use. It is also cataloged as a Herschel 400 object and known as Caldwell 22. (Image taken by The Kingsley School with Faulkes Telescope North operated by Las Cumbres Observatory Global Telescope Network. Used with permission.)

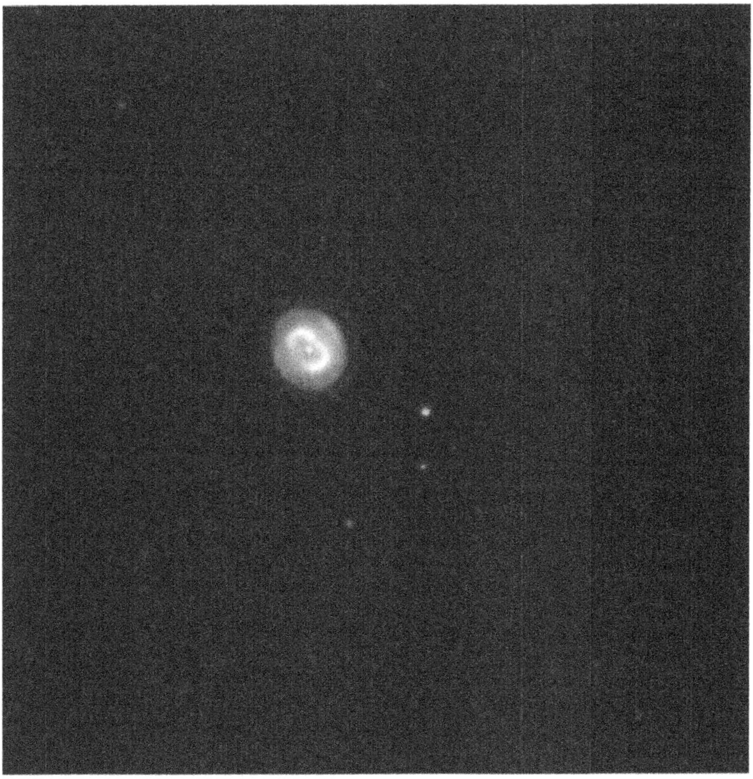

Fig. 8.99 NGC 7662

NGC 7822 Cepheus

RA: 23h 57m
Dec: +68.22
Size: 65×40′
Magnitude: 10.8
Filters: Hα, NB, RGB

This is a huge nebula in Cepheus that lies about 3,000 light years away from Earth and contains one of the hottest main sequence stars known. The principal illuminating star BD+66 1673 is an O-type blue supergiant with a surface temperature of 50,000°. The cloud itself curves around this central star—part of a star cluster known as Berkeley 59, and is diffuse in nature yet crossed by dark lanes of dust and non-illuminated gas, forming several "elephant trunk" formations, and one surprise is that part of the cloud is actually a supernova remnant, indicating that one of the stars of the central cluster has already exploded despite the young age of the group. (Image taken by Martin Griffiths at the Griffon Educational Observatory, Andalucia, Spain.)

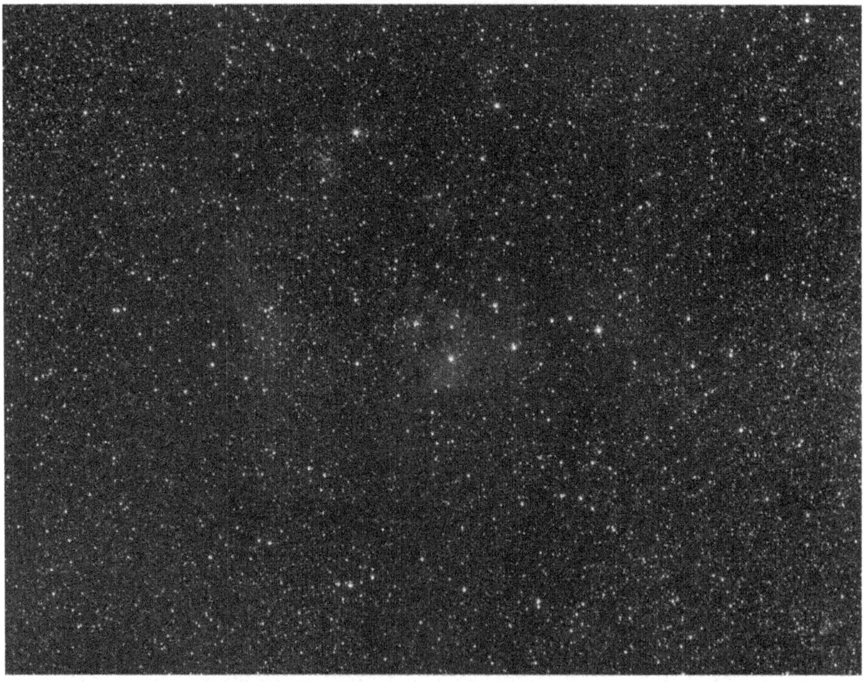

Fig. 8.100 NGC 7822

Further Reading

Allinson M (2005) Star clusters and how to observe them. Springer, New York
Bushnall D (2005) Observing the deep sky. Crowood Press, Marlborough
Chawick S, Cooper I (2012) Imaging the southern Sky. Springer, New York
Coe S (2007) Nebulae and how to observe them. Springer, New York
Gilmour J (2002) The practical astronomers deep sky companion. Springer, New York
Griffiths M (2012) Planetary Nebulae and how to observe them. Springer, New York
Privett G (2007) Creating and enhancing digital astro images. Springer, New York

Glossary of Astronomical Terms

absolute magnitude The magnitude that a star would possess it if were placed at a distance of 10 parsecs from Earth. In this way, absolute magnitude provides a direct comparison of the brightness of stars. The apparent magnitude of a star is based upon its luminosity and its distance. If all stars were placed at the same distance then their apparent magnitudes would only be dependent on their luminosities. Thus, absolute magnitudes are true indicators of the amount of light each star emits.

absorption Absorption is a property of atomic elements where they absorb a photon of light of a particular wavelength, resulting in the electron(s) within the atom either jumping to a higher orbit in the atom (excitation) or leaving the atom altogether, a process known as ionization. This leads to the development of a dark line in the spectrum of a star or other body at the specific energy or wavelength of the absorbed photon.

accretion An accumulation of dust and gas into larger bodies such as stars, planets and moons, or as discs around existing bodies.

albedo A measure of the reflectivity of an object, which is expressed as the ratio of the amount of light reflected by an object to that of the amount of light incident upon it. A value of 1 represents a perfectly reflecting (white) surface, while a value of zero represents a perfectly absorbing (black) surface. Some typical albedos are: Earth—0.39; The Moon—0.07; Venus—0.59.

aphelion The point in an orbit around the Sun at which an object is at its greatest distance from the Sun (opposite of perihelion).

apoapsis The point in an orbit when a planet is farthest from any body other than the Sun or Earth.

apogee Similar to aphelion, the point in an orbit when a body orbiting Earth, (e.g., the Moon or an artificial satellite.) is farthest from Earth (opposite of perigee).

M. Griffiths, *Choosing and Using Astronomical Filters*, The Patrick Moore Practical Astronomy Series, DOI 10.1007/978-1-4939-1044-1,
© Springer Science+Business Media New York 2015

arc minute A measure of angular separation, or one sixtieth of a degree.

arc second Another measure of angular separation, one sixtieth of an arc minute (1/3,600th of a degree).

ascending node The point in the orbit of an object when it crosses the ecliptic (or celestial equator) while moving south to north.

asteroids (also "planetoids") These are rocky bodies, the vast majority of which orbit the Sun between Mars and Jupiter. It is thought that there must be around 100,000 in all. The largest asteroid is Ceres, which has a diameter of 579 miles. The smallest detected asteroids have diameters of several hundred feet. Together with comets and meteoroids, asteroids make up the minor bodies of the Solar System. They are considered to be the leftover planetesimals from the formation of our Solar System. The gravitational pull of Jupiter is thought to have stopped the members of the Asteroid Belt from forming a planet.

astronomical unit (au) This is the mean distance from Earth to the Sun, i.e., 149, 597, 870 km.

aurora A glow in Earth's ionosphere caused by the interaction between Earth's magnetic field and charged particles from the Sun (the solar wind). This gives rise to the "northern lights," or Aurora Borealis, in the northern hemisphere, and the Aurora Australis in the southern hemisphere.

Baader Astro solar film A neutral density film that reduces the intensity of sunlight by 99.99 %, thus allowing direct viewing through an appropriate telescope.

Bessell filters The generally used UBVRI photometric system of color filtration applied to CCD photography.

binary star A system of two stars orbiting around a common center of mass due to their mutual gravity. Binary stars are twins in the sense that they formed together out of the same interstellar cloud.

blue Moon The second full Moon in a calendar month, or the third full Moon in a season containing four.

broadband filter A filter that is generally used to reduce light pollution as it transmits the wavelengths of light for Hα, OIII and Hβ but stops the transmission of light wavelengths inimical to sodium and mercury vapor streetlights.

Caldwell catalog A catalog of 110 objects constructed by the British amateur astronomer Sir Patrick Moore and based on the famous Messier catalog by the eighteenth-century French observer Charles Messier. The Caldwell catalog is named after Moore's surname—which was the hyphenated Caldwell-Moore. It contains objects from the NGC and IC catalogs and covers both southern and northern celestial hemispheres.

celestial equator The projection of Earth's equator upon the celestial sphere.

celestial poles The projection of Earth's poles onto the celestial sphere.

celestial sphere The projection of space onto the night sky, an imaginary hollow sphere of infinite radius surrounding Earth but centered on the observer (first postulated by Ptolemy). It is the basis of sky charts and the celestial coordinate system. The coordinate system most commonly used is right ascension and declination. The sphere itself is split up into arbitrary areas known as constellations.

chromosphere The layer between the photosphere and the corona in the atmosphere of the Sun, or any other star, mainly composed of excited hydrogen atoms. In an Hα telescope the chromosphere appears to have a myriad of bright points across the solar disc, a phenomenon known as the chromospheric network.

coma (1) The dust and gas surrounding the nucleus of a comet. (2) A defect in an optical system that gives rise to a blurred, pear-shaped, comet-like image.

comet An icy object in independent orbit around the Sun, smaller than a planet and usually presenting a highly elliptical orbit extending out to beyond Jupiter.

conjunction When two bodies appear to close together in the sky, i.e., they have the same right ascension. Mercury and Venus are said to be at superior conjunction when they are behind the Sun, and at inferior conjunction when they are in front of it. The outer planets are simply said to be in conjunction when they pass behind the Sun.

constellation An arbitrary grouping of stars that form a pattern. The sky is divided into 88 constellations. These vary in size and shape from Hydra, the Sea Monster, which is the largest at 1,303 square degrees, to Crux, the Cross, which is the smallest at 68 square degrees.

corona The outer layer, and hottest part, of the Sun's atmosphere.

coronagraph A special telescope that blocks light from the Sun's disc, thus creating an artificial eclipse, in order to study the Sun's atmosphere.

cosmic ray An extremely fast, energetic and relativistic (high-speed) charged particle.

cosmos The universe. The word is derived from the Greek word for "everything."

crater A depression in the lunar or planetary surface caused by an impact from a large meteor or asteroid. Generally circular in appearance and occasionally marked with a central peak and collapsed walls.

culmination An object is said to culminate when it reaches its highest point in the sky. For northern observers, this occurs when the object is due south. For southern observers when it is due north.

declination A system for measuring the altitude of a celestial object, expressed as degrees north, or south, of the celestial equator. Angles are positive if a point is north of the celestial equator, and negative if south. It is used, in conjunction with right ascension, to locate celestial objects.

descending node The point in the orbit of an object when it crosses the ecliptic while traveling north to south.

digital camera This can be (1) the single lens reflex camera (SLR), which instead of having a standard film inside now relies on an imaging chip to capture the scene in the same manner as a video camera or (2) a charge coupled device (CCD) camera. It can also refer to any compact digital camera that uses chip technology and to differentiate it from the larger SLR types known as DSLR.

direct (prograde) motion (1) Rotation or orbital motion in an counterclockwise direction when viewed from the north pole of the Sun (i.e., in the same sense as Earth); the opposite of retrograde. (2) The east–west motion of the planets, relative to the background of stars, as seen from Earth.

DMK camera A camera that uses digital technology to capture image files in the form of a movie that can then be downloaded and stacked in an appropriate software program such as *Registax*. DMK cameras are used for lunar, solar and planetary imaging.

dwarf star A star that lies on the main sequence and is too small to be classified as a giant star or a supergiant star. For example, the Sun is a yellow dwarf star.

eccentricity The eccentricity of an ellipse (orbit) is the ratio of the distance between its foci and the major axis. The greater the eccentricity, the more 'flattened' is the ellipse.

eclipse A chance alignment of the Sun and any other celestial object, or two other celestial objects in which one body blocks the light of the Sun, or other body, from the other. In effect, the outer object moves through the shadow of the inner object.

ecliptic The apparent path the Sun (and, approximately, that of the planets) as seen against the stars. Since the plane of Earth's equator is inclined at 23.5° to that of its orbit, the ecliptic is inclined to the celestial equator by the same angle. The ecliptic intersects the celestial equator at the two equinoxes.

Ellerman bombs Microflares in the solar chromosphere associated with magnetic field reconnections, where two opposing streams of ionized material collide with a brief flare of light and energy. A small solar flare.

elongation The angular distance between the Sun and any other Solar System body, usually Earth, expressed in degrees. The term greatest elongation is applied to the inner planets, Mercury and Venus. It is the maximum elongation from the Sun. At greatest elongation, the planet will appear at 50 % phase.

emerging flux region An area on the Sun where a magnetic dipole, or flux tube, is surfacing on the disc and can produce a bipolar sunspot group.

ephemeral regions Limited energy magnetic dipoles with lifetimes of about a day that contain no sunspots. Ephemeral regions can develop anywhere on the Sun, but are more common at mid and lower solar latitudes.

equatorial mount A telescope mount designed so that the two axes, which support it, are aligned, one to the polar axis and the other to Earth's equator. Once an object is centered in the telescope's field of view, only the polar axis need be adjusted to keep the object in view. If the polar axis is driven at sidereal rate, it will counteract the rotation of Earth, keeping the object (except the Moon) stationary in the field of view.

equinox This is the time when the Sun crosses the celestial equator. There are two equinoxes: vernal (spring), around March 21, and autumnal (autumn), around September 23. On these dates, day and night are equal. Actual dates and times vary due to Earth's precession.

faculae Unusually bright spots, or patches, on the Sun's surface. They precede the appearance of sunspots and can remain for some months afterwards.

fibrils Fine structure in sunspot areas associated with spicules and solar activity in the chromosphere.

filament A strand of (relatively) cool gas suspended over the Sun (or star) by magnetic fields. These appear dark against the disc of the Sun. A filament on the limb of the Sun seen in emission against the dark sky is called a solar prominence.

galaxy Vast star system containing thousands of billions of stars, dust and gas, held together by gravity. Galaxies are the basic building blocks of the universe. There are three main types: elliptical, spiral and barred, named after their appearance.

Galilean moons Jupiter's four largest moons: Io, Europa, Ganymede and Callisto. First discovered by Galileo.

geosynchronous orbit Sometimes known as a geostationary orbit, in which a satellite's orbital velocity is matched to the rotational velocity of the planet, and as such, a geostationary satellite would appear to be stationary relative to Earth.

globular cluster A spherical cluster of older stars, often found in galaxies.

granulation The mottled, orange peel appearance of the Sun's surface, caused by convection within the Sun.

Gun Griz photometric system A photometric calibration system for professional use that is referenced with known stars of particular spectral character and brightness.

heliocentric Sun-centered system of cosmology.

hypersensitize The process of treating a photographic film with hydrogen- or nitrogen-forming gas to render the emulsion more sensitive to light and to reduce reciprocity failure with long exposures.

inclination (1) The angle between the orbital plane of the orbit of a planet and the ecliptic. (2) The angle between the orbital plane of a satellite and the equatorial plane of the body it orbits.

inferior conjunction When Mercury, or Venus, are directly between the Sun and Earth.

inferior planets These are the planets Mercury and Venus. They are called inferior planets because their orbits lie between that of Earth and the Sun.

light year The distance traveled by light in 1 year, equal to 9.4607^{12} km.

limb The outer edge of the disc of a celestial body.

luminence layer The image taken by a CCD camera through a hydrogen alpha, SII or Ca II filter, which is then added to an BVR image to gain maximum input from the astrophysical image.

luminosity Absolute brightness. The total energy radiated into space, per second, by a celestial object such as a star.

lunation The period between successive new Moons.

magnetosphere The region of space where a planet's magnetic field dominates that of the solar wind.

magnitude The degree of brightness of a celestial body designated on a numerical scale, on which the brightest star is magnitude −1.4 and the faintest star visible to the unaided eye is magnitude 6. A decrease of one unit represents an increase in apparent brightness by a factor of 2.512. Apparent magnitude of a star is the brightness as we see it from Earth, while absolute magnitude is a measure of its intrinsic luminosity. Lower numbers represent brighter objects.

mare Area on the lunar surface that was once thought to be a sea of water (*mare* is Latin for "sea"). Any open surface on a planet that is a lava plain.

meteor Also known as a "shooting star" or "falling star," this is a bright streak of light in the sky caused by a meteorite as it burns up in Earth's atmosphere.

meteorite A rock of extraterrestrial origin, found on Earth.

minor planets Another term for asteroids.

moon A naturally occurring satellite, or relatively large body, orbiting a planet. The Moon, usually written with an initial cap, refers to Earth's Moon.

Mylar filter A solar filter that allows less than 1 % transmission of light through a metalized filter to enable safe solar viewing in white light.

nebula A term used to describe a celestial object that has a fuzzy, or nebulous, appearance (from the Latin for "cloud"), such as gas, or dust, clouds.

nebula filters Generally wide bandpass filters or light pollution filters that allow the passage of Hα, OIII and Hβ wavelengths through to a camera, optical system or CCD camera.

nova An existing star that suddenly increases its brightness by more than ten magnitudes and then slowly fades. Novae are generally associated with binary stars in which one of the stars is a white dwarf in close proximity to the primary star. The primary star sheds gas to the white dwarf, which allows build up on the surface until pressure and temperature ensure a huge thermonuclear detonation.

occultation This is when one celestial body passes in front of, and obscures, another.

open cluster A group of young stars, possibly bound together by gravity, that formed together.

opposition A planet is said to be "in opposition" when it appears opposite the Sun in the sky. For the outer planets, this is generally the closest they come to Earth, hence when they are most easily visible.

optical binary A pair of stars that happen to lie close to one another on the celestial sphere because of a chance alignment. They are not physically associated with one another and lie at vastly different distances. Optical binaries are also known as visual binaries.

orbit The path of one body around another due to the influence of gravity.

parallax The angular difference in apparent direction of an object seen from two different viewpoints.

Parsec A unit for expressing large distances. It is the distance at which a star would have a parallax of 1 arc second, equal to 3.2616 light years or 206,265 astronomical units (au).

penumbra Means, literally, "dim light." It most often refers to the outer shadow cast during eclipses, and defines the region of shadow that gives rise to a partial eclipse. It is also the lighter area surrounding the central region of a sunspot.

periapsis The point in an orbit closest to a body other than the Sun or Earth.

perigee The point in its orbit where the Moon, or a planet, is closest to Earth.

perihelion The point in its orbit when an object is closest to the Sun.

perturb To cause a celestial body to deviate from its predicted orbit, usually under the gravitational influence of another celestial object.

photosphere The visible surface of the Sun.

plage Bright region in the Sun's chromosphere.

planisphere An aid to locating stars and constellations in the night sky. It consists of two discs, one with the entire night sky and the other, which covers the first, having a window through which a portion of the sky can be seen. The second disc is set according to the date and time.

precession Circular motion about the axis of rotation of a body; fixed with respect to the stars. Earth is a giant gyroscope whose axis passes through the North and South poles, and this axis precesses with a period of 27,700 years.

prominence A cloud, or plume, of hot, luminous gas in the solar chromosphere. It appears bright when seen against the cool blackness of space. When they are in silhouette against the disc prominences are known as filaments. They are mainly composed of hydrogen, helium and calcium.

quadrature When a superior planet such as Jupiter or Saturn is at right angles to the Sun, as seen from Earth.

quasars Compact, extragalactic objects at extreme distances, which are highly luminous. They are thought to be active galactic nuclei. The name is an acronym for quasi-stellar radio source. A quasar is very similar to a QSO (quasi-stellar object) but gives out radio waves also.

radiant The part of the sky from which a particular meteor stream appears to come. Meteor showers are usually named after the constellation in which the radiant originates.

red giant A spectral type K or M star nearing the end of its life and having a low surface temperature and large diameter, e.g., Betelgeuse in Orion.

red shift The lengthening of the wavelength of electromagnetic radiation caused by relative motion between source and observer. Spectral lines are red-shifted from distant galaxies, indicating that the galaxies are moving away from us due to the expansion of the universe.

resolution The amount of small detail visible in an image (usually telescopic); low resolution shows only large features, high resolution shows many small details.

retrograde Rotation of a planet, or orbit, opposite to that normally seen.

right ascension (RA) The angular distance, measured eastwards, from the vernal equinox. It is one of the ordinates used to reference objects on the celestial sphere and is the equivalent to a longitude reference on Earth. There are 24 h of right ascension within 360°, so 1 h is equivalent to 15°. Together with declination, it represents the most commonly used coordinate system in modern astronomy.

semi-major axis The semi-major axis of an ellipse (e.g., a planetary orbit) is half the length of the major axis, which is a segment of a line passing through the foci of the ellipse with end points on the ellipse itself. The semi-major axis of a planetary orbit is also the average distance from the planet to its primary.

sidereal month The 27.32166-day period of the Moon's orbit.

sidereal time Star time; the hour angle of the vernal equinox. Time measured with respect to the fixed stars rather than the Sun.

solar continuum filter A green light filter transmitting light wavelengths centered at 510 nm, rendering a visible green image of the sun. Such filters are used in conjunction with either a Herschel wedge or Baader astro filters.

solar cycle The 11-year variation in sunspot activity.

solar flare A sudden, short-lived burst of energy on the Sun's surface, lasting from minutes to hours.

solar wind A stream of charged particles emitted from the Sun that travels into space along lines of magnetic flux.

solstice This is the time when the Sun reaches its most northerly or southerly point (around June 21 and December 22, respectively.). It marks the beginning of summer and winter in the northern hemisphere and the opposite in the southern hemisphere.

spectral classification A method of classifying stars based upon the appearance of absorption lines in their spectra.

star cluster A loose association of stars within the Milky Way. Examples are the Pleiades (Seven Sisters) and Hyades clusters.

sunspot A cooler region of the Sun's photosphere (which, thus, appears dark) seen as a spot on the Sun's disc. Sunspots are caused by concentrations of magnetic flux, typically occurring in groups or clusters. The number of sunspots varies according to the Sun's 11-year cycle. More sunspots are seen at the maxima of solar cycles, with few being observed during the minima between.

superior conjunction This is when Mercury, or Venus, is behind the Sun.

superior planets Also known as the outer planets. These are the planets beyond Earth's orbit. They are, in order, Mars; Jupiter, Saturn, Uranus, Neptune, and the dwarf planet Pluto.

supernova An exploding star.

terminator The boundary between day and night regions of the Moon's, or a planet's, disc.

ToUcam A small webcam that fits in the eyepiece holder of a telescope to gain a direct video image of an astronomical object. Manufactured by Phillips.

transit The apparent journey of Mercury or Venus across the Sun's disc, or of a planet's moon across the disc of its parent.

UBVRI The colored filter photometric system generally employed by amateur astronomers and systematized by Michael Bessell in the 1990s.

umbra From the Latin for shade, it is the shadow area defining a total eclipse. Or, an umbra is the dark central region of a sunspot.

unsharp masking A photographic and image reduction technique that allows the stacking of many images to gain increased detail and resolution in an astronomical object.

variable star Any star whose brightness or magnitude varies with time. The variations can be intrinsic, because of internal processes, or extrinsic, due to eclipses, dust and other phenomena. Variations can also be irregular or periodic.

white dwarf A whitish star, of up to 1.4 solar masses, and about the size of Earth with consequential very high density, characterized by a high surface temperature and low brightness.

Wratten filters Colored glass filters with a range across the visible spectrum from red to blue that enable the blocking of particular longpass wavelengths of light in order to see more detail on planetary and lunar surfaces. The filters have a range of colors and are numbered according to accepted standards.

zenith The point on the celestial sphere directly above an observer, or the highest point in the sky reached by a celestial body.

zenithal hourly rate (ZHR) This is the number of meteors per hour, for a particular stream, that is estimated will be seen under favorable seeing conditions if the radiant were directly overhead the observer. Usually the actual figure is less than this.

Zirin class The different active or quiescent features of prominences in the solar chromosphere developed by Harold Zirin.

zodiac The apparent path, in the sky, followed by the Sun, Moon and most planets, lying within 10° of the celestial equator. Ancient astrologers (nothing to do with modern astronomy!) divided the zodiac into 12 groups, the signs of the zodiac, though there are actually 13 astronomical constellations that lie on the zodiac, since the Sun passes through Ophiuchus each December. Ophiuchus is not recognized by astrologers.

zodiacal light A faint glow from light scattered off interplanetary dust in the plane of the ecliptic.

Further Reading

Astrophotography

Buick T, Pugh P (2011) How to photograph the Moon and planets with your digital camera. Springer, New York
Covington M (2007) Digital SLR astrophotography. Cambridge University Press, Cambridge, UK
Gendler R (2013) Lessons from the masters: current concepts in image processing. Springer, New York
Hall A (2013) Getting started: long exposure astrophotography. Createspace Independent Publishing Platform, USA
Howell S (2006) Handbook of CCD astronomy. Cambridge University Press, Cambridge, UK
Hubbell G (2013) Scientific astrophotography. Springer, New York
Ireland RS (2009) Photoshop astronomy. Wilman-Bell, Richmond, USA
Privett G (2007) Creating and enhancing digital astro images. Springer, New York
Seip S (2007) Digital astrophotography: a guide to capturing the Cosmos. Rocky Nook Publishing. Sebastapol, USA
Wodaski R (2002) The new CCD astronomy. New Astronomy Press, Maple Valley, USA

Astronomical Instrumentation

Mobberly M (1999) Astronomical equipment for amateurs. Springer, New York
Mobberly M (2004) The new amateur astronomer. Amazon media. Springer, New York
Pugh P (2007) Observing the Sun with Coronado telescopes. Springer, New York

M. Griffiths, *Choosing and Using Astronomical Filters*, The Patrick Moore Practical Astronomy Series, DOI 10.1007/978-1-4939-1044-1,

© Springer Science+Business Media New York 2015

Observing

Allinson M (2005) Star clusters and how to observe them. Springer, New York
Bushnall D (2005) Observing the deep sky. Crowood Press, UK
Chawick S, Cooper I (2012) Imaging the southern sky. Springer, New York
Coe S (2007) Nebulae and how to observe them. Springer, New York
Gilmour J (2002) The practical astronomers deep sky companion. Springer, New York
Grego P (2005) The Moon and how to observe it. Springer, New York
Griffiths M (2012) Planetary Nebulae and how to observe them. Springer, New York
Jenkins J (2009) The Sun and how to observe it. Springer, New York
Kier R (2009) 100 best astrophotography targets. Springer, New York
Luginbuhl C, Skiff B (1998) Observing handbook and catalogue of deep sky objects. Cambridge University Press, Cambridge, UK
Rumistrzewicz S (2010) A visual astronomer's photographic guide to the deep sky. Springer, New York

Index

M. Griffiths, *Choosing and Using Astronomical Filters*, The Patrick Moore Practical Astronomy Series, DOI 10.1007/978-1-4939-1044-1,
© Springer Science+Business Media New York 2015

Lightning Source UK Ltd.
Milton Keynes UK
UKOW06f0250230916

283578UK00003B/34/P

9 781493 910434